Current Cancer Research

Series Editor
Wafik El-Deiry

More information about this series at http://www.springer.com/series/7892

Philip W. Hinds • Nelson E. Brown

Editors

D-type Cyclins and Cancer

 Springer

Editors
Philip W. Hinds
Tufts University School of Medicine
Boston, MA, USA

Nelson E. Brown
University of Talca Medical School
Talca, Chile

ISSN 2199-2584   ISSN 2199-2592   (electronic)
Current Cancer Research
ISBN 978-3-319-87797-6   ISBN 978-3-319-64451-6   (eBook)
DOI 10.1007/978-3-319-64451-6

Printed on acid-free paper

This Springer imprint is published by Springer Nature
The registered company is Springer International Publishing AG
The registered company address is: Gewerbestrasse 11, 6330 Cham, Switzerland

Preface

The devotion of this volume to the role of D-type cyclins in tumorigenesis is reflective of the central role that these cell-cycle regulatory proteins play in the generation and maintenance of a vast array of human malignancies. This family of proteins has been and continues to be the subject of a voluminous primary and review literature over 25 years, and cyclin D1 in particular stands among auspicious company (e.g., ras, myc, p53) as one of the most common central drivers of cancer. This collection of reviews and opinions is meant to present a historical perspective of the work that led to the discovery and biological insight into the key roles for the D-type cyclins in normal cellular processes and disease and the clinical targeting of its enzymatic partners cdk4 and cdk6. Perhaps more importantly, several additional chapters provide summaries of important, recent advances in understanding a much more complex regulation of cyclin D1 production and function that was initially appreciated, as well as roles for the D-type cyclins that extend well beyond their function as cdk activators and/or extend the understanding of cyclin D/cdk function outside of its commonly accepted role as a driver of the G1-to-S phase transition in cycling cells.

As elegantly articulated in the first chapter from Sherr and Sicinski, the origin story of the D-type cyclins, identified over two and a half decades ago, includes a rather unusual element of "instant" appreciation for the central role of these proteins in both cell-cycle control and cancer. That is, D-type cyclins were initially discovered by Matsushime (working in the Sherr lab) as mediators of cell proliferation signals emanating from outside the cell and transduced by growth factor receptors on the cell surface, immediately suggesting a role for D-type cyclins in connecting nuclear events to the sensing of the cell's environment. Contemporaneously, the mammalian D-type cyclins were shown to complement yeast mutants deficient for their own cyclins (Xiong and coworkers), giving biological support to functional inferences drawn from interspecies and interfamily protein homology, and importantly, cyclin D1 was also identified as the likely target of chromosomal rearrangement in parathyroid tumors (Motokura and coworkers). Thus, within a short span of time, researchers using the most cutting-edge functional assays of the period identified the same proteins as likely mammalian G1-phase cyclins that were critical for transducing growth signals from growth factor receptors and that were in at least one instance (and subsequently appreciated to be much more widely observed) the direct target of genetic oncogenic events. The sum of these discoveries was the generation of a hypothesis that continues to be tested and explored to this day – a central regulator of cellular decisions to proliferate (or not) that acts as an obligate partner of specific members of the cyclin-dependent kinase family is commonly exploited by tumor cells ostensibly to remove the dependence of the cell on upstream signaling events, thus freeing the cell to divide at will, producing neoplasms in a wide variety of tissues.

Despite this rapid start to the appreciation of D-type cyclins as key instigators of normal and aberrant cellular proliferation, the study of exactly how they and their partner cdks effect these biological responses has spanned the subsequent two decades and continues to this day. Initial functional studies in cells, and importantly in a wide variety of genetically engineered mice, strongly supported the perceived roles of D-type cyclins as decision-makers in the cell cycle in normal cells and in tumors. These studies are summarized in the chapter from Kalaszczynska and Ciemerych and importantly point out the repeated finding that deficiencies in D-type cyclins produce deficiencies in the development of normal tissues. This in turn provided clues to the proclivity of tumors to deregulate D-type cyclins: perhaps the lack of control engendered by amplifications, rearrangements, and mutations of this family of proteins locks certain cells in a more embryonic state, providing the seeds for subsequent generation of tumors.

Several following chapters then explore the function and regulation of D-type cyclins in molecular detail – from the finding that isoforms of cyclin D1 generated posttranscriptionally that evade normal regulatory controls may be key to oncogenesis in many tumors (Diehl and Knudsen) to an expansion of the understanding of functions of D-type cyclins that extend well beyond their canonical roles as regulators of the retinoblastoma protein (pRB), the best understood substrate of D cyclin/cdk4(6) complexes, and the function that is presumed to be key to the success of cdk4/6 inhibitors that are currently in clinical use. Examples of these functions extend to direct roles in transcriptional control (with or without partner cdks; DiSante et al.) and to a role in nutrient sensing and metabolism that may dictate important physiological responses to loss of D cyclin/cdk function and that may be key to future combinatorial therapies designed to push malignant cells into a state that favors their elimination versus stasis (Valenzuela and Brown).

The volume ends with a provocative chapter from Dowdy summarizing work that redefines the fundamental role of the cyclin D/cdk4(6) complex as a modifier of pRB function. Here, the author expounds on recently published evidence that intriguingly suggests that cyclin D/cdk4(6) complexes may function to produce a sort of "pRB code" by modifying one and only one of some 14 individual phospho-acceptor serines or threonines in pRB, such that a cell entering G1 from a resting state may express any or all of 14 different subspecies of pRB distinguished by phosphorylation on one of these residues. Importantly, these cyclin D-stimulated modifications of pRB seem to favor the association of pRB with its downstream targets, rather than disrupt them, as the canonical model of cyclin D1 function now posits. The biological implications of this study are multifold and include the hypothesis that the oncogenic function of D-type cyclins may not involve pRB inactivation per se, but rather may focus on one of the other functions described above, such as metabolic control or kinase-dependent transcriptional regulation (depending on cell type?) that favors exit from a resting state (or failure to enter a resting state, such as that accompanying the process of differentiation). Clearly, the story of the D-type cyclins is far from fully written, and the work described in this volume, produced in numerous labs over many years, will continue to help reveal and refine the normal and tumorigenic functions of these proteins. This in turn will ultimately improve our understanding of cell-cycle regulation, tissue formation in development, and, importantly, the best way to employ exciting new therapeutics targeting these proteins in human cancers.

Boston, MA, USA   Philip W. Hinds
Talca, Chile   Nelson E. Brown

Contents

Chapter 1
The D-Type Cyclins: A Historical Perspective

Charles J. Sherr and Peter Sicinski

Abstract D-type cyclins integrate mitogen-dependent signals to enforce progression through the first gap phase (G1) of the cell division cycle. In simplest terms, three mammalian D-type cyclins (D1, D2, and D3), induced in a cell lineage-specific fashion in response to extracellular signals, interact with two cyclin-dependent kinases (CDK4 and CDK6) to form holoenzyme complexes that phosphorylate the retinoblastoma protein (RB). In turn, RB phosphorylation, reinforced by other CDKs expressed later in G1 phase, inactivates the suppressive effects of RB on transcription factors that induce genes required for DNA replication. All steps in the life history of individual D-type cyclins, including their transcriptional induction, translation, assembly with CDK4 and CDK6, and their rapid turnover via ubiquitin-mediated proteolysis, are governed by mitogen signaling. Hence, progression through the G1 phase of the mammalian cell cycle is tied to extracellular signals that ultimately influence cell division. Analysis of phenotypes of mice lacking D cyclins has highlighted their individual and combinatorial lineage-specific activities during mammalian development. The genes encoding D-type cyclins and their dependent kinases, CDK4 and CDK6, are proto-oncogenes implicated in many forms of cancer. Genetic or biochemical disruption of cyclin D-dependent CDK signaling can restrain cancer development and progression. Here, we highlight the founding discoveries.

Keywords Cell cycle • G1-phase progression • CDK4 • CDK6 • Retinoblastoma protein (RB) • CDKN2A • p16^{INK4a} • RB pathway • Cancer • Palbociclib

C.J. Sherr (✉)
Howard Hughes Medical Institute, Department of Tumor Cell Biology,
St. Jude Children's Research Hospital, Memphis, TN 38105, USA
e-mail: sherr@stjude.org

P. Sicinski
Department of Genetics, Harvard Medical School, and Department of Cancer Biology,
Dana Farber Cancer Institute, Boston, MA 02215, USA

© Springer International Publishing AG 2018   1
P.W. Hinds, N.E. Brown (eds.), *D-type Cyclins and Cancer*,
Current Cancer Research, DOI 10.1007/978-3-319-64451-6_1

The discovery of the D-type cyclins and their biochemical and genetic characterization more than two decades ago provided early insights about how extracellular signals are integrated with the core cell cycle machinery to drive cell proliferation. Conversely, the realization that unrestrained expression of D-type cyclins and their allosterically regulated cyclin-dependent kinases, CDK4 and CDK6, occurs frequently in tumor cells pointed to mechanisms by which cancer cells abandon environmental controls to acquire greater autonomy and an increased capacity for cellular self-renewal. Here, we trace early work on the D-type cyclins first begun with their discovery in 1991, concentrating on foundational findings primarily made throughout the 1990s. We discuss the history of these gathered insights, some predicted and some not, which provided a basic understanding of how D-type cyclins function and that galvanized later work in the field, including the recent clinical use of drugs targeting cyclin D-dependent kinases in cancer.

1.1   The Core Cell Cycle Machinery

Cyclins, first named for their cyclic expression during the different phases of the cell division cycle, bind to, and allosterically activate, cyclin-dependent kinases (CDKs) whose phosphorylation of critical substrates governs cell cycle progression. The pivotal realization that a cyclin-dependent kinase controls mitotic entry and exit stemmed from genetic experiments performed with both budding yeast (*Saccharomyces cerevisiae*) and fission yeast (*Schizosaccharomyces pombe*) and from convergent biochemical investigations in frogs, starfish, and clams that originated in the 1970s and 1980s (reviewed in [1–3]). A collision of multiple lines of evidence established in the late 1980s that a mitotic cyclin (cyclin B) assembled with a single CDK [p34^{Cdc28} in budding yeast, p34^{cdc2} in fission yeast, and henceforth CDK1] to form an active holoenzyme that regulates mitosis in all eukaryotes. Extending the paradigm, the periodic expression of other evolutionarily conserved cyclins governs progression through other phases of the cell cycle (Fig. 1.1).

The genes that control the orderly alternating periodicities of chromosomal DNA synthesis (S phase) and mitosis (M phase) in proliferating cells encode various cyclins, CDKs, and other key regulators of their activities. Together, the encoded proteins represent the core cell cycle machinery that functions as an S-phase/M-phase oscillator in a largely cell autonomous fashion. Indeed, under normal physiological circumstances, cells that commit to enter S phase are driven by feed-forward mechanisms to complete the cell cycle. Additional checkpoint and feedback controls exerted at various points in the cycle help to guarantee that one essential process (e.g., faithful DNA replication) is completed before another begins (e.g., mitosis). Other checkpoint controls exist to ensure that DNA damage is repaired before mother cells segregate chromosomes to daughter cells or that correct assembly of the mitotic spindle occurs before chromosome separation at anaphase [5]. As might be expected, the components of the core cell cycle machinery are highly conserved evolutionarily among all eukaryotes. With respect to CDK1, for example, both

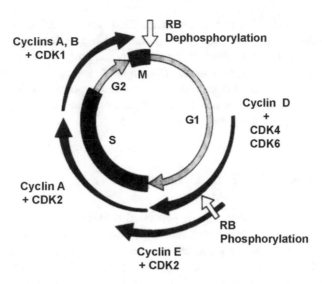

Fig. 1.1 Composition of mammalian cyclin-CDK complexes during the cell cycle. Cyclin D- and E-dependent kinases contribute to RB phosphorylation late in G1 to facilitate the commitment of cells to enter S phase. CDK2 and CDK1 activities mediated by cyclins A and B drive cells through S, G2, and M phase, after which hyperphosphorylated RB is returned to its hypophosphorylated state (Adapted from Fig. 1 in [4] with permission from Elsevier)

fission yeast *cdc2* and human *CDK1* can rescue *Cdc28* deficiency in budding yeast despite the remarkable divergence of these species [6, 7]. In short, the core cell cycle machinery is dedicated to allow accurate replication of DNA and to faithfully distribute the duplicated genetic material to two daughter cells.

In metazoans, the earliest, rapid embryonic cell divisions are represented by repeated S phases and M phases that depend upon maternally derived proteins that are not initially rate limiting. However, once zygotic transcription begins, S phase and M phase become separated by two gap phases, the first (G1) between M phase and S phase and the second (G2) between S phase and M phase (Fig. 1.1).

Progression through G1 phase is particularly sensitive to extracellular environmental signals, triggered by nutrients, mitogens, antiproliferative cytokines, cell-to-cell contacts, and other spatial cues (reviewed in [4]). It is at this time that cells sense their environments to "decide" whether to withdraw from the cell cycle or continue through G1 toward S phase and eventual mitosis.

In unicellular eukaryotes such as budding yeast, the activity of a single CDK (p34^{Cdc28}/CDK1) is regulated by distinct classes of G1-, S-, and M-phase cyclins that are expressed periodically during different cell cycle intervals (reviewed in [8]). The decision to divide ("START" in *S. cerevisiae*) occurs during G1 phase and depends largely on the availability of extracellular nutrients that, when plentiful, stimulate unicellular growth (increased cell size) and determine the activity of G1 cyclin (Cln)-dependent p34^{Cdc28} to trigger S-phase entry and subsequent cell division. Alternatively, haploid cells in G1 phase can undergo arrest and conjugate to form diploids, a program triggered by α mating factors [9, 10]. In response to these pheromones, a protein (Far1) associates with Cln-Cdc28 complexes to inhibit their kinase activities and prevent entry into S phase. Notably, cells that pass START are resistant to both nutrient deprivation and α-factors and progress through the remainder of the cell cycle. These findings in yeast highlight the general principle that the irreversible commitment of eukaryotic cells to enter S phase enables the core cell cycle machinery to drive downstream events in a manner that is largely insulated from physiological environmental stimuli.

1.2   Discovery of D-Type Cyclins

Mammalian cells respond to signals generated by numerous environmental cues and, like START in budding yeast, reach a so-called restriction point in late G1 phase, where they irreversibly commit to enter S phase, lose their dependency on mitogens, and complete the cell division cycle [11]. By 1990, investigators in the cell cycle field assumed that mammalian cells would also express G1 cyclins analogous to the three identified Cln proteins in yeast and that these would regulate the kinase activity of CDK1 or perhaps other novel CDKs.

By introducing a cDNA library from human glioblastoma cells into conditionally *CLN*-deficient *S. cerevisiae*, Xiong and coworkers [12] identified a single cDNA that rescued yeast cell proliferation under restrictive growth conditions. Nucleotide sequencing revealed that this cloned *c*DNA, designated cyclin D1, had homology to known cyclins A and B and yeast Clns but was clearly distinct. At the same time, Matsushime et al. [13] identified a new "cyclin-like" gene (originally designated Cyl1 as per referees' doubts that it was a bona fide cyclin) by the use of a subtraction hybridization screen to clone genes from a mouse macrophage cell line stimulated to synchronously enter G1 phase by colony-stimulating factor-1 (CSF-1). A fortuitous meeting of the principal investigators followed by exchange of predicted amino acid sequences between these two groups revealed that human cyclin D1 and mouse Cyl1 were virtually identical. In turn, an antiserum prepared to the recombinant mouse 36 kDa Cyl1 protein specifically reacted with human cyclin D1. Unexpectedly, however, immunoprecipitates containing Cyl1 did not contain CDK1 and failed to support an associated kinase activity that phosphorylated the then canonical CDK1 substrate, histone H1. Instead, using metabolically radiolabeled mouse macrophages, an unrelated 34 kDa protein was coprecipitated using the antiserum to Cyl1, raising the suspicion that it might represent a novel CDK (see below). Also unanticipated, the Cyl1 protein, while induced in G1 phase, did not periodically "cycle" in proliferating macrophages, but was continually synthesized as long as CSF-1 stimulation was maintained. Finally, using a Cyl1 probe, two closely related cyclins, Cyl2 and Cyl3 (now designated cyclins D2 and D3, respectively), were cloned and found to be expressed in mitogen-stimulated mouse T lymphocytes that lacked Cyl1 expression [13, 14]. Hence, unlike the three Cln proteins of budding yeast, the three mouse D-type cyclins were not co-expressed together but instead seemed to be synthesized in a cell lineage-specific manner in response to stimulation by different mitogens.

Adding an important dimension to these findings, Motokura et al. [15] identified a candidate oncogene (*PRAD1*) on human chromosome 11q13 that was clonally rearranged and expressed after chromosomal inversion [inv(11)(p15;q13)] under the control of the parathyroid hormone promoter. Cloning of the PRAD1 cDNA revealed its homology to then known cyclins and documented its expression in various human, mouse, and bovine tissues. When the PRAD1 protein was added to interphase clam embryo lysates lacking endogenous cyclins, it activated the histone H1 kinase activity of recovered p34^{cdc2} (CDK1). Comparison of the human PRAD1 nucleotide sequence with that of human cyclin D1 showed them to be identical. Most telling, identification of PRAD1 strongly implied that cyclin D1 had proto-oncogenic activity. Indeed, as Motokura and coworkers recognized, the same region of human chromosome 11q13 included the previously designated *BCL1* oncogene known to be amplified in breast and squamous cell carcinomas and to be targeted by translocations [t11;14 (q13;q32)] in mantle cell lymphomas.

Taken together, these papers led to the following conclusions and speculations:

* At least three different D-type cyclins are expressed in mammalian cells, presumably in a cell lineage-dependent manner in response to mitogen stimulation. Given the nonuniform distribution of cyclin D1, D2, and D3 protein expression in different mitogen-responsive cell lines, no single D-type cyclin is likely to be essential for the cell cycle.
* Cyclin D1 is evolutionarily conserved across mammalian species and can rescue proliferation of conditionally G1 cyclin-deficient yeast, and its associated kinase activity can phosphorylate and activate histone H1 in clam extracts. However, and perhaps paradoxically, no such CDK activity could be demonstrated in mouse cyclin D1 (Cyl1) immunoprecipitates, which contained an unrelated coprecipitating 34 kDa protein that was speculated to be a novel CDK.
* Cyclin D1 is a proto-oncogene that likely contributes to several forms of human cancer.

1.3 Expanding Roles for Mammalian Cyclins and CDKs

The earliest reports describing D-type cyclins were accompanied by a flurry of publications by investigators in the cell cycle field who were actively working to identify novel CDKs and cyclins. Cyclin A, already known to form active complexes with CDK1 prior to mitosis, was revealed to bind to a second 33 kDa protein, designated CDK2, that shares 65% sequence identity with CDK1 but is activated earlier in the cell cycle [16–18] (Fig. 1.1). Yeast complementation, independently used to clone cyclin D1 [19], yielded two additional cyclins designated C and E [19, 20]. Cyclin C was later implicated to associate with CDK3 to facilitate exit of quiescent cells from G_0 [21]; moreover, its interaction with CDK8 is now understood to play a role in regulating transcription [22]. Cyclin E accumulates periodically near the G1/S transition and, like cyclin A, associates with CDK2 to drive S-phase entry [23, 24] (Fig. 1.1).

Additional protein kinases that were cloned based on somewhat lesser sequence homology to CDK1 included CDK3 that, like CDK2, could complement *Cdc28/CDK1* mutants in *S. cerevisiae*, as well as at least eight other potential family members that could not [25]. By convention, CDK1-related kinases that bound to cyclins were named in the order of their discovery. In contrast, those candidate kinases that had no known cyclin partners took their names from the amino acid sequences in a highly conserved region, the so-called PSTAIRE motif (single amino acid code) of CDK1. However, despite rapid progress, the identity of cyclin D-dependent CDKs remained unknown.

1.4   CDK4 and CDK6 and Their Roles in RB Phosphorylation

Using a biochemical approach in lieu of yeast complementation, Matsushime and coworkers [26] attempted to utilize glutathione-S-transferase fusion proteins containing each of the three recombinant D-type cyclins to "pull down" radiolabeled products transcribed and translated in vitro from cDNA templates encoding putative CDKs. One of these CDK candidates designated PSKJ3 had been previously cloned from human HeLa cells based on homology to mixed oligonucleotide probes representing highly conserved regions of mammalian kinases [27]. All three D-type cyclins physically interacted with PSKJ3. This finding was surprising because PSKJ3 showed less than 50% amino acid homology to CDKs 1, 2, and 3; lacked the ability to complement Cdc28 deficiency in budding yeast [25]; did not bind to *S. pombe* Suc1 beads that were routinely used to precipitate CDK1 (e.g., see [15]); and lacked a conserved PSTAIRE motif, instead containing a highly divergent PV/ISTVRE sequence.

Previous reports had indicated that the retinoblastoma protein (RB) is cyclically phosphorylated during the cell division cycle (reviewed in [28]). RB is dephosphorylated as cells exit mitosis, and the hypophosphorylated form detected in early G1 phase becomes hyperphosphorylated in late G1 and remains so throughout progression through the cell cycle until cells complete mitosis [29–32]. The fact that the hypophosphorylated G1-phase form of RB acts as a potent tumor suppressor was highlighted by its binding to DNA tumor virus oncoproteins that cancel its function [33–36]. Moreover, the amino acid motifs surrounding mapped sites of RB phosphorylation were typical of sites targeted by CDKs [37], and RB phosphorylation by known CDKs was soon documented [38–40]. Thus, G1 cyclin-CDK complexes were likely to play decisive roles in this process.

Based on collaborations undertaken between the Sherr and Livingston laboratories, Matsushime and coworkers [26] attempted to use a glutathione-S-transferase-RB fusion protein as a cyclin D-PSKJ3 substrate. Baculovirus vector-mediated coexpression of each of the recombinant D-type cyclins with PSKJ3 in Sf9 insect cells allowed the resulting complexes to bind to GST-RB beads and phosphorylate the RB protein (Fig. 1.2). Thus, the ability of D-type cyclins to bind to and allosterically

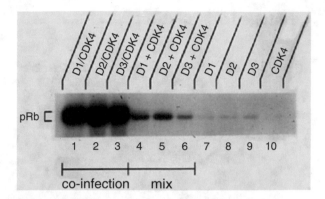

Fig. 1.2 RB kinase activity reconstituted from baculovirus-encoded proteins in insect Sf9 cells. Cells coinfected with baculovirus vectors encoding D-type cyclins and CDK4 (lanes 1–3) were lysed and assayed for RB kinase activity using a GST-RB fusion protein as substrate. Lysates from cells infected with vectors encoding different D-type cyclins were mixed with lysates expressing CDK4 and similarly assayed (lanes 4–6). Lysates containing the individual components (lanes 7–10) lacked kinase activity (The data are composited from Fig. 4 [41] and Fig. 3 [42] with permission from Elsevier and Cold Spring Harbor Laboratory Press, respectively)

activate the kinase activity of PSKJ3 established that the latter was a bona fide CDK, subsequently designated CDK4 [26]. Continuing studies revealed that complexes assembled in insect cells that contained CDK4 and D-type cyclins, but not cyclins A, B, or E, bound directly to and phosphorylated RB but not histone H1 or several other tested substrates [41]. In turn, physical complexes between different D-type cyclins and RB were detected in mammalian cells although at the time, opposing models speculated about the significance of these interactions, and in particular whether D-type cyclins functioned upstream or downstream of RB [42, 43].

Following cyclin D1 induction in metabolically labeled mouse macrophages stimulated with CSF-1 to enter the cell cycle from quiescence, increasing quantities of CDK4 were seen to bind to immunoprecipitated cyclin D1 as cells approached the G1/S boundary (Fig. 1.3). Although it initially proved difficult to immunoprecipitate enzymatically active CDK4 complexes from mammalian cells, the development of appropriate detergent lysis conditions and use of suitable non-inhibitory monoclonal antibodies revealed CDK4-dependent RB kinase activity in proliferating cells [45] (Fig. 1.3c). Cyclin D1-CDK4 activity was not detected in quiescent (G0) cells, whereas macrophages or fibroblasts expressed the active CDK4 kinase when stimulated to enter the cell cycle with CSF-1 or serum, respectively. After mitogen stimulation, the rate of appearance of CDK4 kinase activity during G1 phase lagged significantly behind cyclin D1 induction (Fig. 1.3a), correlating with formation of cyclin D1-CDK4 complexes in mid-G1 (Fig. 1.3b) and with steady increases in their associated RB kinase activity as cells approached S phase (Fig. 1.3c). As expected from earlier work [23, 24, 39], phosphorylation of RB by cyclin E- and A-driven CDK2 was detected near the G1/S transition, increased as cells progressed through the cycle, and was maintained until cells exited mitosis

8

C.J. Sherr and P. Sicinski

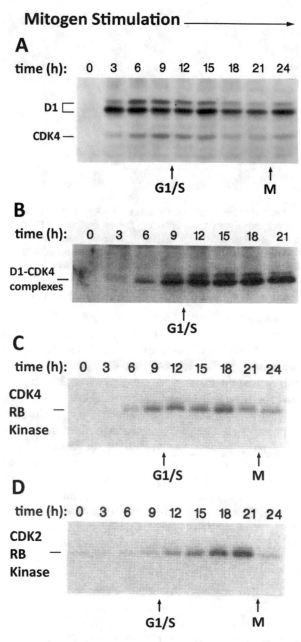

Fig. 1.3 Kinetics of cyclin D synthesis, assembly with CDK4, and activation of RB kinase activity in macrophages entering G1 phase from quiescence (G_0). (**a**) Macrophages made quiescent by CSF-1 withdrawal reentered the cell cycle synchronously following CSF-1 restimulation. Cells metabolically labeled with [^{35}S] methionine during each 3 h interval after CSF-1 stimulation were lysed with detergent at the indicated times, and radiolabeled proteins precipitated with antiserum to cyclin D1 were detected by autoradiography after electrophoretic separation on a denaturing polyacrylamide gel. Cyclin D1 was first detected after 3 h of mitogen exposure and continued to

[46] (Fig. 1.3d). Intriguingly, enforced ectopic expression of either cyclin D1 or D3 together with CDK4 in rodent fibroblasts made quiescent by serum starvation failed to induce any CDK4 kinase activity unless the cells were stimulated to reenter the cell cycle [45]. Hence, upstream regulators, dependent upon mitogen-induced signals, were needed to govern formation of the active holoenzymes.

Later studies by Meyerson and Harlow [47] indicated that yet another kinase, CDK6, was activated by the D-type cyclins to phosphorylate RB. CDK6 exhibits 71% amino acid identity with CDK4 and contains a PLSTIRE motif. Like D-type cyclins, CDK4 and CDK6 are differentially expressed in different cell types, sometimes but not always in an overlapping manner. Association of cyclin D2 with CDK6 in phytohemagglutinin (PHA)-stimulated T cells (which express cyclins D2 and D3, but not cyclin D1) was observed as early as 6 h after PHA stimulation, but RB phosphorylation was not detected until cells entered mid-G1 phase 6 h later. Later in G1, cyclin D3 assembled into active complexes with CDK6, and cells entered S phase about 30 h after PHA stimulation. The independent temporal regulation of cyclins D2 and D3 observed in primary T lymphocytes was thought to highlight distinct molecular roles of these proteins in cell cycle progression [48]. With current knowledge, we realize that the varying kinetics of cyclin D2 and D3 induction more likely reflect stereotypic responses to sequential signal transduction events that rely on the T-cell receptor and interleukin-2 receptor, respectively. Despite these temporal differences, it was clear that the timing of CDK6 activation in T cells, as well as CDK4 in macrophages and fibroblasts, occurs well after cells have entered G1 phase from G_0 but precedes the activation of cyclin E-CDK2 and cyclin A-CDK2 which does not ensue until cells are near the G1/S transition.

As noted above, the appearance of active cyclin D-dependent kinase activities after entry into G1 phase from G_0 requires transcriptional induction of D-type cyclins and their mitogen-regulated assembly with CDKs, processes which require many hours. However, in asynchronously dividing cells, the situation is rather different, because cyclin D and CDK4/CDK6 synthesis are maintained as long as

---

Fig. 1.3 (continued) be synthesized as long as CSF-1 stimulation was continued. The more slowly migrating species in the cyclin D1 doublet is a phosphorylated form. An even more rapidly migrating coprecipitating protein was later shown to be CDK4. The G1/S transition, determined by analysis of DNA content and BrdU labeling, occurred between 9 and 12 h after stimulation (*arrow*). (**b**) Macrophage cell lysates collected at 3 h intervals as in panel **a** were precipitated with antiserum to cyclin D1, and proteins recovered on protein A-conjugated Sepharose beads were separated on a denaturing polyacrylamide gel, transferred to a nylon membrane, and immunoblotted with an antiserum to CDK4. D1-CDK4 complexes were first detected at 6 h, and cells entered S phase between 9 and 12 h (*arrow*). (**c**) Cyclin D1-CDK4 complexes recovered as in panel **b** were incubated in the presence of $\Upsilon[^{32}P]$-ATP and a recombinant RB substrate, and the radiolabeled products were electrophoretically separated on a denaturing gel and detected by autoradiography. Intense RB kinase activity was detected prior to the G1/S transition (*arrow*). (**d**) Complexes precipitated with antiserum to CDK2 were assayed for RB kinase activity as in panel **c** (Data are adapted from Fig. 4 (panel **a**) [13] with permission from Elsevier and from Fig. 2 (panels **b**, **c**, and **d**) [45] with permission from the American Society of Microbiology)

Cell cycle entry and exit

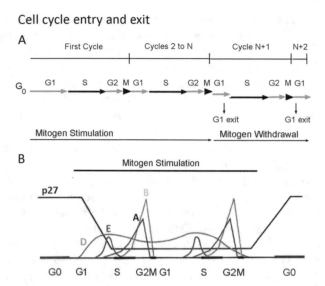

Fig. 1.4 Cell cycle entry and exit. (**a**) In cells entering the division cycle from quiescence (G_0) in response to mitogenic stimulation, the D-type cyclins must be induced and assembled with CDKs before the complexes phosphorylate RB. However, during subsequent cycles, maintenance of cyclin D-dependent complexes results in the subsequent shortening of G1 phase in ensuing cycles (2 to N). For asynchronously dividing cells completing mitosis, mitogen withdrawal leads to prompt G1-phase arrest in cycle N + 1; however, cells that have already entered S phase complete cycle N + 1 and withdraw from the cycle in the ensuing G1 phase. (**b**) The schematic illustration documents periodic expression of various cyclins and the levels of p27[Kip1] in cycling and quiescent (Go) cells (Panel **a** is unpublished; panel **b** is reproduced from Fig. 1 [44], with permission from the American Association for the Advancement of Science)

mitogenic stimulation continues. Under these circumstances, the need for cyclin D induction (but not the requirement for CDK4/CDK6 activity) is negated as cells reenter G1 phase, the duration of which following mitosis is contracted (Fig. 1.4).

1.5 How do Mitogens Regulate Cyclin D-Dependent Kinases?

Of the three D-type cyclins, most early studies were performed with cyclin D1. Ras signaling promotes transcription of the cyclin D1 gene via a kinase cascade that depends upon sequential RAF-MEK-ERK signaling [49–52]. Examples of other diverse mitogenic signals that conjoin to induce cyclin D1 include Wnt/β-catenin in the colon [53, 54] and estrogen signaling, particularly in the breast [55–57]. Because D-type cyclin mRNAs and their encoded proteins are highly unstable ([13, 26] and see below), withdrawal of mitogens not only results in termination of cyclin D transcription but also rapidly leads to cyclin D proteolysis [58]. Hence, cells entering the cycle in response to only brief mitogen stimulation, or cycling cells that have completed mitosis, do not enter S phase unless mitogenic stimulation is persistent (Fig. 1.4).

Assembly of cyclin D1-CDK4 complexes also depends upon the RAF-MEK-ERK signaling pathway [59], although the underlying mechanisms remain ill defined. Cyclin D1-CDK4/D1-CDK6 complexes have molecular masses of ~150 kDa that supersede those of cyclin D1 (36 kDa) + CDK4 (34 kDa). Surprisingly, CDK inhibitors of the Cip/Kip family including p21^{Cip1} and p27^{Kip1} bind to cyclin D-CDK4/D-CDK6 complexes during G1 phase [60, 61] and co-purify in higher order complexes with the active kinases [62–66], suggesting that Cip/Kip proteins may function as "assembly factors." In agreement with this concept, in fibroblasts doubly null for p21^{Cip1} and p27^{Kip1} expression, assembly of cyclin D-CDK complexes and their nuclear accumulation are drastically reduced, whereas re-expression of Cip/Kip proteins in these cells fully restores cyclin D-CDK activity [67]. Genetic inactivation of *Kip1* largely rescues effects of cyclin D1 inactivation during mouse development (see below) further underscoring the ability of cyclin D1 complexes to stoichiometrically "titrate" p27^{Kip1} in living animals [68, 69].

Why don't bound Cip/Kip proteins inhibit cyclin D-dependent holoenzymes? One intriguing possibility is that the inhibitory activity of p21^{Cip1} and p27^{Kip1} is negated by their phosphorylation by growth factor regulated non-receptor tyrosine kinases [70–73]. For example, phosphorylation of p27^{Kip1} on tyrosine-88 within the p27^{Kip1} kinase inhibitory domain results in ejection of the CDK inhibitor from the ATP binding pocket of CDK2, while other portions of p27^{Kip1} remain bound to active CDK2-cyclin A [70]. Thus, mitogen-stimulated tyrosine phosphorylation of Cip/Kip proteins may allow them to trigger cyclin D-CDK assembly without inactivating the kinase activity.

Sequestration of CDK inhibitors of the Cip/Kip family also points to a non-catalytic role of the cyclin D-CDK4/D-CDK6 complexes that facilitates activation of cyclin E-CDK2 as cyclin D-CDK complexes accumulate (reviewed in [74]) (Fig. 1.5). In quiescent cells, the levels of p27^{Kip1} in particular are elevated and must be reduced for cells to enter S phase and subsequently cycle (Fig. 1.4b). The titration of CDK inhibitors sets a dependency of cyclin E-CDK2 on mitogen-dependent assembly of cyclin D-CDK complexes, which is absent in cells engineered to lack Cip/Kip proteins [67]. Once its enzymatic function is unencumbered, cyclin E-CDK2 enhances its own activity by phosphorylating and triggering the proteolytic degradation of p27^{Kip1} as cells approach the G1/S transition [75, 76], and residual p27Kip1 remains bound to cyclin D-CDK complexes in proliferating cells [60, 61] (Fig. 1.5). The concerted activities of the G1-phase cyclin-dependent kinases disrupt the interactions of RB (and of other RB family proteins, p130 and p107) with various E2F transcription factors, allowing their coordinated regulation of some hundreds of genes whose activities are required for the initiation and onset of DNA synthesis (reviewed in [77, 78]).

Complexes of E2Fs with different RB family members play roles in both activating and repressing E2F target genes. For example, during G1 phase, the RB family member p130 in association with E2F4 represses E2F-responsive genes, whereas RB itself sequesters activating E2Fs-1, E2Fs-2, and E2Fs-3 to inhibit their

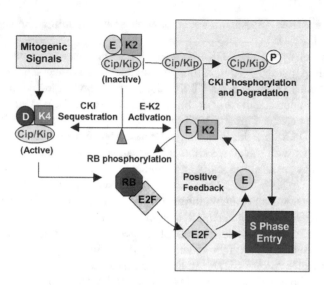

Fig. 1.5 Regulation of the G1/S transition. Mitogenic signals promote the assembly of active cyclin D-dependent kinases containing either CDK4 (or CDK6, not shown) and a Cip (p21Cip1) or Kip (p27Kip1) protein. Sequestration of Cip/Kip proteins lowers the overall inhibitory threshold and facilitates activation of cyclin E-CDK2. The cyclin D- and E-dependent kinases contribute sequentially to RB phosphorylation, canceling RB's ability to repress E2F family members that trigger S-phase entry. E2F responsive genes include cyclins E and A. Cyclin E-CDK2 further reinforces RB phosphorylation to drive S-phase entry and phosphorylates p27 to trigger its proteolysis. The degradation of Cip/Kip proteins and induction of cyclins by E2F (*highlighted by background shading*) contribute to mitogen independence and the irreversibility of the transition (Data are taken from Fig. 1 [74] with permission from Cold Spring Harbor Laboratory Press)

transcriptional activity [79, 80]. The inhibitory effects of RB family proteins on transcription are reversed when both RB and p130 are phosphorylated by G1 cyclin-dependent CDKs as cells exit G1 and enter S phase. Notably, targets of activating E2F transcription factors include cyclins E and A, providing a positive feed-forward loop that further reinforces the activation of CDK2 and contributes to the irreversible commitment of cells to enter S phase (Fig. 1.5).

In the continued presence of mitogenic signals, assembled cyclin D-CDK complexes enter the nucleus and undergo activating phosphorylation on a single "T-loop" threonine residue (Thr-172) by a CDK-activating enzyme (CAK), itself composed of cyclin H and CDK7 (formerly called MO15) [81–83]. CAK activity appears to be constitutively active throughout the cell cycle. Because neither the D-type cyclins nor their associated CDKs contain canonical nuclear import signals, nuclear localization of the assembled holoenzymes may depend upon their association with Cip/Kip proteins [64, 67]. The import of assembled cyclin D-CDK4/D-CDK6 holoenzymes into the nucleus increases during G1 phase and, in asynchronously proliferating cells, is maximal near the G1/S transition [84]. During S phase, cyclin D1 is exported from the nucleus, and its rate of degradation is increased [85, 86].

The turnover of D-type cyclins is regulated by phosphorylation at a single threonine residue near the C-terminus (Thr-286 in cyclin D1). Phosphorylation of Thr-286 generates a "phosphodegron" that triggers cyclin D1 ubiquitination and its rapid proteasomal degradation [58]. The GSK-3β kinase, which is negatively regulated by the phosphatidyl-inositol 3-kinase (PI3K) signaling pathway, is responsible for Thr-286 phosphorylation [85]. Therefore, active mitogen-dependent signaling through the PI3K-AKT-GSK-3β cascade inhibits GSK-3β and stabilizes CDK4-bound cyclin D1. Even under stimulatory conditions, the half-life of cyclin D1 is short ($t_{1/2}$ = 30 min), but mitogen withdrawal or chemical inhibition of PI3K reduces its half-life to less than 8 min. By contrast, mutation of cyclin D1 Thr-286 to alanine markedly increases cyclin D1 half-life to >4 h and results in its nuclear retention in S phase [85, 86]. Hence, phosphorylation of cyclin D1 on Thr-286 is required for its CRM-dependent transport to the cytoplasm during S phase and its subsequent degradation [86].

In summary, many steps in the life history of cellular cyclin D1 (and by inference, D2 and D3) are mitogen dependent. These include cyclin D1 transcription, assembly with CDK4, nuclear import and export, and ubiquitin-mediated proteolysis.

1.6   Lineage-Dependent Functions of D-Type Cyclins in Mice

Genetic inactivation of D-type cyclins in mice, either alone or in combination, provided key insights about how they function in development [summarized in Table 1.1]. In agreement with observations that certain tissues preferentially express only one of the three D cyclins, a common theme is that mice lacking individual D cyclin family members exhibit corresponding focal defects. Mice lacking cyclin D1 exhibit severe retinal hypoplasia and profound defects in mammary gland lobulo-alveolar development during pregnancy, consistent with the fact that cyclin D1 is strongly expressed in the retina and breast [87, 88] (Fig. 1.6). A genetic knock-in of cyclin D2 into the cyclin D1 locus largely rescued the phenotype of D1 deficiency, indicating that these two D-type cyclins are functionally interchangeable [103]. Other defects in *Ccnd1*-null mice, including growth retardation and abnormal clasping reflexes (Fig. 1.6e), implicate cyclin D1 function in the central nervous system and perhaps more globally in other tissues in which other D-type cyclins are also expressed. In contrast, cyclin D2-null mice have defects in ovarian granulosa cells leading to infertility, as well as testicular hypoplasia and more subtle defects in cerebellar and B-lymphocyte development [91–94] (Fig. 1.7). The latter deficiencies reflect the fact that cyclins D2 and D1 are co-expressed in early postnatal cerebellar granule neuron progenitors (Fig. 1.7a), whereas cyclins D2 and D3 are both synthesized in lymphocytes. D2-null mice also develop hyperglycemia that proceeds to frank diabetes due to defective pancreatic β-cell expansion [95], consistent with observations that CDK4 inactivation also leads to insulin-dependent diabetes [105, 106, 108]. Cyclin D3-null mice show hematopoietic deficiencies primarily

Table 1.1 Phenotypes of mice lacking D-type cyclins, CDK4 and CDK6

Disrupted gene(s)	Survival	Pathology	References
D1	Viable	Small body size, hypoplastic retinas, postnatal photoreceptor degeneration, defective lobulo-alveolar breast development during pregnancy, abnormal neurological clasping reflexes, resistance to Ras/Her2 oncogene-induced breast cancer	[87–90]
D2	Viable	Defective ovarian granulosa cell expansion, hypoplastic testes, abnormal postnatal cerebellar development with reduced numbers of granule neurons, impaired adult neurogenesis, impaired B-lymphocyte proliferation, glucose intolerance progressing to diabetes	[91–96]
D3	Viable	Defective thymic T-cell maturation from CD4/CD8 double-negative to CD4/CD8 double-positive cells with cytokine-independent defects in pre-T-cell receptor signaling, impaired B-cell and germinal center development, defective responses to granulocyte colony-stimulating factor and reduced neutrophil numbers, resistance to T-cell leukemogenesis	[97–100]
D1, D2	Viable but early postnatal death	Retarded growth, hypoplastic retinas, defective postnatal cerebellar development, and impaired coordination	[101]
D1, D3	Early postnatal death, but a few survivors to 2 months	Neurological abnormality, failure to thrive, hypoplastic retinas	[101]
D2, D3	Embryonic lethality before E18.5	Megaloblastic anemia	[101]
D1, D2, D3	Death before E17.5	Severe hematopoietic defects affecting stem and progenitor cells. Transplanted fetal liver cells cannot reconstitute lymphoid or myeloid function. Defects in heart development and anemia. MEFs show greatly reduced susceptibility to transformation by Ras plus Myc, E1A, or dominant-negative p53	[102]
D2 → D1[a]	Viable	Nearly complete rescue of cyclin D1 deficiency	[103]
D2 → D3[a]	Viable	No rescue of cyclin D3 deficiency	[104]
CDK4	Viable	Small body size, male and female infertility, deficiency in pancreatic beta-cell regeneration, and insulin-deficient diabetes. Quiescent fibroblasts exhibit delay in S-phase entry after mitogen stimulation associated with increased binding of $p27^{Kip1}$ to cyclin E/A-CDK2	[105, 106]
CDK6	Viable	Mild hematopoietic impairment	[107]
CDK4, CDK6	Late embryonic death	Severe anemia	[107]

[a]Replacement of D2 coding sequences for the indicated D-type cyclin

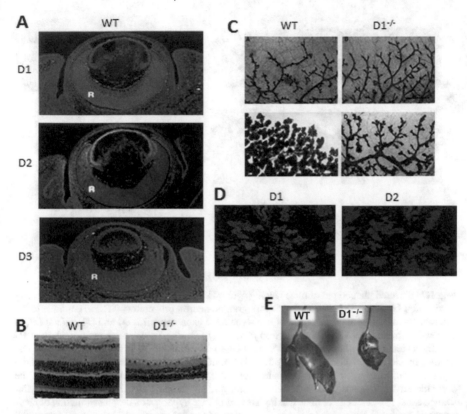

Fig. 1.6 Phenotypes of mice lacking cyclin D1. (**a**) In situ hybridization detection of cyclin D1, D2 and D3 transcripts in embryonic retinas (R) of wild-type mice. Cyclin D1 is the major D cyclin expressed in this tissue. (**b**) Sections of retinas stained with hematoxylin and eosin. Note retinal hypoplasia in cyclin D1$^{-/-}$ animals. (**c**) Whole-mount appearance of mammary epithelium in virgin female mice (*top panels*) and after delivery of pups (*lower panels*). Lobulo-alveolar development is reduced following pregnancy in cyclin D1$^{-/-}$ females. (**d**) In situ detection of cyclin D1 and D2 RNAs in mammary glands of pregnant wild-type females. (**e**) When raised by their tails, cyclin D1-null mice display an abnormal "leg-clasping" reflex (Panel **a** is reproduced from [68], and panels **b**, **c**, and **e** are taken from [88], all with permission from Elsevier. (**d**) (previously unpublished) was kindly provided by Drs. Susan B. Parker and Stephen J. Elledge (with permission from S.J.E.))

affecting both T and B lymphocytes, erythrocyte macrocytosis, and diminished responses to granulocyte colony-stimulating factor (G-CSF) [97–99, 109]. Surprisingly, replacement of cyclin D3 with cyclin D2 did not rescue cyclin D3 deficiency, arguing that these two D cyclins have different functions that supersede their tissue-specific expression [104].

If the tissue-specific manifestations of single cyclin D elimination determine the observed developmental deficiencies, mice lacking more than one D cyclin should exhibit additive phenotypes. However, mice engineered to express only a single D-type cyclin developed normally until late gestation; here, tissue-specific expression

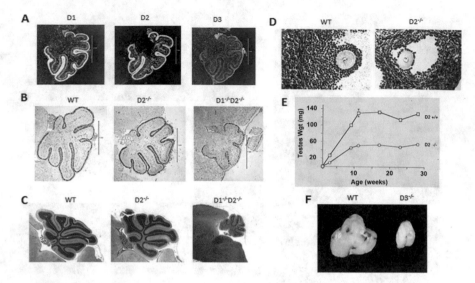

Fig. 1.7 Selected phenotypes of mice lacking cyclin D2 or D3. Cerebellar development primarily occurs after birth. (**a**) At postnatal day 5 (P5), granule neuron progenitors (GNPs) confined to the external granule layer of the cerebella of wild-type mice express cyclins D1 and D2, but not D3, as detected by in situ RNA hybridization (*top panels*). (**b**) Bromodeoxyuridine incorporation at P5 documents reduced proliferation of GNPs in the cerebella of mice lacking both cyclins D1 and D2. (**c**) By P16, GNPs have exited the cell cycle and have migrated into the internal granule layer of the organ (as confirmed by their strong hematoxylin staining within this anatomic compartment). The outer molecular layer has become devoid of GNPs and appears as a zone cleared of hematoxylin-stained cells. The cerebella of mice lacking D1 and D2 are hypoplastic, despite formation of characteristic folia. (**d**) Hematoxylin and eosin staining of ovaries reveals hypoplastic granulosa cell layers in ovarian follicles of cyclin D2-null females. (**e**) Cyclin D2-null males exhibit testicular hypoplasia. (**f**) The thymi of adult cyclin D3-null mice are hypoplastic when compared to age-matched wild-type mice (Panels **a**, **b**, and **c** are reproduced from [101] with permission from Cold Spring Harbor Press. Panel **d** is reproduced from [91] with permission from Nature Publishing Group. Panel **f** is reproduced from [97] with permission from Elsevier. Panel **e** was not previously published)

of D cyclins was lost, and mutant embryos ubiquitously expressed the one remaining D cyclin [101]. Hence, the functions of the three D cyclins are largely exchangeable at this stage of development. However, later in life, tissues lacking two of three D-type cyclins failed to upregulate the remaining family member and exhibited compound deficiencies. For example, mice lacking cyclins D1 and D2 have severe defects in cerebellar development and die postnatally (Fig. 1.7a, b), whereas those with inactivated genes encoding both D2 and D3 die before embryonic day (E) 18.5 with severe anemia [101]. Remarkably, mice lacking all three D cyclins live until E16.5 and die with megaloblastic anemia and anatomical heart defects [102]. The latter animals exhibit other severe hematopoietic defects affecting the number and proliferation of stem and progenitor cells (Fig. 1.8). In this setting, CDK4 and CDK6 lack detectable kinase activity, and cell cycle progression relies critically on the activity of the "downstream" kinase CDK2. Mouse embryo fibroblasts (MEFs) established from embryos lacking the three D-type cyclins could still be propagated in culture, implying that

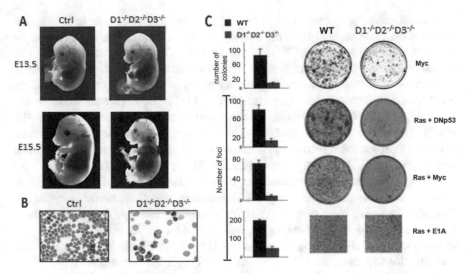

Fig. 1.8 Phenotype of mice and cells lacking all three D-type cyclins. (**a**) Appearance of embryos at embryonic days 13.5 (E13.5) and 15.5 (E15.5). Cyclin $D1^{-/-} D2^{-/-} D3^{-/-}$ embryos do not present overt abnormalities at E13.5, but they appear pale at 15.5. (**b**) Wright-Giemsa stained peripheral blood smears from wild-type and cyclin D-deficient embryos documents a paucity of blood cells in cyclin $D1^{-/-} D2^{-/-} D3^{-/-}$ animals. (**c**) Reduced sensitivity of cyclin $D1^{-/-} D2^{-/-} D3^{-/-}$ mouse embryonic fibroblasts to transformation by the indicated oncogenes, including Myc, Ras, dominant-negative (DN) p53, and adenovirus E1A. Bar graphs depict mean numbers of foci (or colonies in case of c-Myc); error bars indicate standard deviation from the mean. Also shown are representative crystal violet-stained monolayers of cells (Reproduced from [102] with permission from Elsevier)

cyclin E−/A-dependent CDK2 can "sense" mitogenic signals in the complete absence of D-type cyclins. However, primary D-type cyclin-null MEFs exhibited reduced susceptibility to transformation by various oncogenes [102] (Fig. 1.8). Remarkably, mice engineered to lack both CDK4 and CDK6 displayed strikingly similar phenotypes to animals lacking all three D cyclins, thereby providing an in vivo validation that these proteins function in the same pathway [107].

Together, these findings reveal that mice lacking one or more D cyclins undergo normal embryogenesis at least until mid-gestation and that focal defects emerging relatively late in fetal development may not affect postnatal viability and seemingly involve only a minority of tissues. In short, the D-type cyclins are dispensable for the workings of the cell cycle per se.

1.7 D-Type Cyclins and Cancer: The top of the Iceberg

The original finding that cyclin D1 was a target of chromosomal inversion in parathyroid adenoma [15] quickly gave way to the realization that *CCND1* was translocated or amplified in many other common adult human cancers. In mantle cell lymphomas, *CCND1* is translocated and placed under control of the immunoglobulin heavy chain locus [110, 111], thereby driving ectopic D1 expression in B cells that normally express only cyclins D2 and D3. In the smaller fraction of these tumors that lack *CCND1* translocations, *CCND2* is rearranged [112]. Amplification of *CCND1* is frequent in squamous carcinomas of the head and neck, esophageal carcinomas, bladder and primary breast cancer, small cell lung cancers, and hepatocellular carcinomas (reviewed in [44, 113]). The amplicons are large, but invariably involve *CCND1* compared with flanking genes, and are associated with cyclin D1 overexpression. Indeed, genome-wide analyses revealed that the cyclin D1 gene represents the second most frequently amplified locus across all human tumor types [114]. Aberrantly elevated cyclin D1 expression is also seen in sarcomas, colorectal tumors, and melanomas, despite little evidence of gene amplification in these tumors. In about 6% of bladder cancers, mutations affecting cyclin D1 threonine-286 have been detected (http://www.cbioportal.org/data_sets.jsp), these being predicted to stabilize cyclin D1 and increase both its nuclear activity and oncogenic potential ([85, 86]; reviewed in [115]). Translocations, overexpression, and mutations affecting the stability of the other D cyclins have also been described, particularly in various human lymphoid malignancies [116–122]. Similarly, soon after its discovery, the gene encoding CDK4 located on chromosome 12q13 was found to be co-amplified together with *MDM2* (encoding the p53 E3 ligase) in sarcomas and gliomas [123–126]. Likewise, the CDK6 gene is amplified in several tumor types [127–130]. These findings established roles for D-type cyclins and their catalytic partners as proto-oncogenes.

The discovery of a highly specific 16 kDa polypeptide inhibitor of CDK4 encoded by the *INK4a* (formally *CDKN2A*) gene [131] helped to firmly establish that CDK4 acts upstream of RB. The p16^{INK4a} protein binds directly to and inhibits the two cyclin D-dependent kinases but no other CDKs, preventing RB phosphorylation and arresting cells in the G1 phase of the cell cycle. Notably, cells lacking functional RB were resistant to p16^{INK4a}-mediated cell cycle arrest, implying that the ability of CDK4 and CDK6 to drive G1-phase progression requires RB [132–134]. A reverse genetic screen independently identified *INK4a* as a gene associated with susceptibility to familial melanoma, and, forecasting what was to come, its immediate study demonstrated its homozygous deletion at high frequencies in cell lines derived from tumors of the lung, breast, brain, bone, skin, bladder, kidney, ovary, and lymphocytes [135]. Remarkably, the small locus (<50 kB) on human chromosome 9p21 that encompasses both *CDKN2A/INK4a* and *CDKN2B/INK4b* is now recognized to encode three closely linked tumor suppressors; these include not only the two inhibitors of CDK4 and CDK6 (p16^{INK4a} [131] and p15^{INK4b} [136]) but also the functionally unrelated ARF tumor suppressor which acts to induce p53 activity and is encoded in part from an *INK4a* alternate reading frame from which ARF gets its name [137, 138]. Inactivating mutations within unique regions of the p16^{INK4a} open reading frame are commonly found in melanoma [139] and in some other tumor types, whereas deletion and silencing of the entire *CDKN2A/B* locus are among the most common events in human cancers (http://www.cbioportal.org; [114]). These findings pointed to the significance of disruption of the "RB pathway," in which p16^{INK4A} and RB function as canonical tumor suppressors and

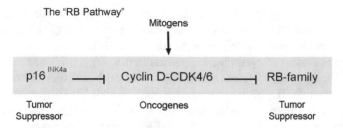

Fig. 1.9 The "RB pathway." Mitogen-activated cyclin D-CDK4/CDK6 complexes phosphorylate RB to help cancel its tumor suppressive function. Activation of p16^{INK4a} by various forms of oncogenic stress inhibits CDK4/CDK6 activity to maintain RB in its hypophosphorylated active form. The RB signaling pathway, highlighted by background shading, is so frequently disrupted in cancer that its inactivation may be necessary for tumor development [28, 44]

undergo mutually exclusive inactivating mutations in many human cancers, whereas D-type cyclins and their associated CDKs act as oncogenes (reviewed in [44, 113, 140]) (Fig. 1.9).

1.8   Drug-Induced Inhibition of Cyclin D-Dependent Kinases in Cancer Treatment

By the mid-1990s, the elucidated biochemical and epistatic genetic relationships between p16^{INK4a}, cyclin D-CDK4/D-CDK6, and RB already provided clear and early proof of principle that potential drug-induced inhibition of cyclin D-dependent CDKs that should mimic the effects of p16^{INKa} might have utility in cancer treatment. Importantly, although the components of cyclin D-dependent kinases proved nonessential for the cell cycle per se, the clinical efficacy of drugs specifically targeting these enzymes might reflect an underlying acquired "addiction" of cancer cells to the effects of disruption of the RB pathway while sparing normal cells that maintain physiological signaling thresholds.

Leaping two decades forward, a drug inhibiting CDK4/CDK6 (palbociclib, Ibrance™) received conditional FDA approval in 2015 for the treatment of estrogen receptor-positive breast cancers [141]. In one trial cohort, patients were required to have cancers with *CCND1* amplification, loss of p16^{INK4a}, or both to be enrolled. FDA approval of the drug was based on dramatic increases in progression-free survival documented in a Phase II clinical trial performed with women who received palbociclib together with the aromatase inhibitor letrozole, versus a control group that received letrozole alone [141]. Palbociclib was well tolerated with more than 87% of patients enrolled in the cohort remaining on study. Neutropenia, unaccompanied by related infections, was the most common side effect. Predicted side effects of long-term targeted therapy may involve glucose intolerance, based on the requirements of pancreatic β-cells for CDK4 [105, 106] and demonstrated effects of p16^{INK4a} on age-dependent islet cell regeneration [142]. If so, such effects should be readily managed by adjusting treatment schedules or by the use of available antidiabetic regimens. In retrospect, the underlying rationale for the use of CDK4/CDK6 inhibitors in estrogen receptor-positive breast cancer is multifold, in the sense that (i) estrogen is a potent driver of cyclin D1 expression in the breast ([57]; recently reviewed in [143]), (ii) deletion of cyclin D1 or CDK4 prevents breast tumorigenesis spurred by activation of the Her2/Ras pathway in mice [89, 144, 145], (iii) cyclin D1-CDK kinase activity is required for breast tumorigenesis [146], and (iv) an acute and global shutdown of cyclin D1 in mice bearing HER2-positive breast cancers selectively halts the proliferation of tumor cells while having no visible effects on nontransformed tissues [147]. Hence, whereas letrozole inhibits estrogen-driven cyclin D1 synthesis in mammary tissues, palbociclib blocks the kinases activated by the cyclin.

Palbociclib (first designated PD-0332991) was synthesized at now defunct Parke-Davis in 2001 by Dave Fry and Peter Toogood [148]. It differed from previously developed and significantly toxic, broader acting CDK inhibitors by its specificity for cyclin D-dependent kinases. However, despite earlier scientific inroads predicting that CDK4/CDK6 inhibition might prove efficacious in cancer treatment, PD-0332991 was not immediately utilized. After acquisition of Parke-Davis/Warner Lambert by Pfizer, Inc., PD-0332991 failed to show much antitumor effect in a Phase 1 trial in 2004, and its clinical development was halted until late 2009. Conceivably, while CDK4 inhibition might have failed as a monotherapy, the judicious combination of a CDK4/CDK6 inhibitory drug together with other agents that block cyclin D synthesis and stability, such as letrozole in ER-positive breast cancer, proved effective. Perhaps, combining MEK or PI3K inhibitors, both of which influence the cellular life history of D-type cyclins (see above), might synergize with CDK4/CDK6 inhibitor treatment in other cancers. Despite the long hiatus in advancing palbociclib through clinical trials, a continuing and expanding interest of many pharmaceutical companies in developing additional CDK4/CDK6 inhibitory drugs, coupled with clinical trials already in progress or planned, highlights how basic investigations of the D-type cyclin-dependent kinases eventually led to practical impact in cancer medicine [149].

The lessons here are many. Frequently, early discoveries predate the available technologies required to exploit them. For CDK4/CDK6 inhibitors, it took a decade before chemists were able to develop a specific inhibitor that proved medicinally acceptable as a soluble, stable, potent, nontoxic, and orally available compound. This is an arena in which industry excels. Even so, the merger of Parke-Davis with Pfizer, and a plethora of competing new agents brought into the Pfizer pipeline, stalled clinical utilization of PD-0332991 for which there may not have been a vocal in-house advocate. Even when tried as a monotherapy in Phase I trials, the drug seemed to lack antitumor activity. Whether chosen by insight or by default, the combined use of palbociclib in letrozole-treated patients with estrogen receptor-positive breast cancer in a well-designed Phase II trial finally revealed its potent activity, leading to FDA breakthrough status and subsequent approval for clinical use. Had there been better communication between basic scientists and the development team, progress might have been accelerated. Future work will determine whether the promise of CDK4/CDK6 inhibitors in breast cancer treatment holds true and whether other drug combinations can be exploited in other cancers.

References

1. Hunt T. Cyclins and their partners: from a simple idea to reality. Semin Cell Biol. 1991;2:213–22.
2. Dorée M, Hunt T. From Cdc2 to Cdk1: when did the cell cycle kinase join its cyclin partner? J Cell Sci. 2002;15:2461–4.
3. Norbury C, Nurse P. Animal cell cycles and their control. Annu Rev Biochem. 1992;61:441–70.
4. Sherr CJ. Mammalian G1 cyclins. Cell. 1993;73:1059–65.
5. Hartwell LH, Weinert TA. Checkpoints: controls that ensure the order of cell cycle events. Science. 1989;246:629–34.
6. Booher R, Beach D. Site-specific mutagenesis of cdc2+, a cell cycle control gene of the fission yeast *Schizosaccharomyces pombe*. Mol Cell Biol. 1986;6:3523–30.
7. Lee MG, Nurse P. Complementation used to clone a human homologue of the fission yeast cell cycle control gene cdc2. Nature. 1987;327:31–5.
8. Reed SI. G1-specific cyclins: in search of an S-phase promoting factor. Trends Genet. 1991;7:95–9.
9. Wittenberg C, Sugimoto K, Reed SI. G1 specific cyclins of *S. cerevisiae*: cell cycle periodicity, regulation by mating pheromone, and association with the p34^{CDC28} protein kinase. Cell. 1990;62:225–37.
10. Chang F, Herskowitz I. Identification of a gene necessary for cell cycle arrest by a negative growth factor of yeast: *FAR1* is an inhibitor of G1 cyclin, *CLN2*. Cell. 1990;63:999–1011.
11. Pardee AB. G1 events and regulation of cell proliferation. Science. 1989;246:603–8.
12. Xiong Y, Connolly T, Futcher B, et al. Human D-type cyclin. Cell. 1991;65:691–9.
13. Matsushime H, Roussel MF, Ashmun RA, et al. Colony-stimulating factor-1 regulates novel cyclins during the G1 phase of the cell cycle. Cell. 1991;65:701–3.
14. Matsushime H, Roussel MF, Sherr CJ. Novel mammalian cyclin (CYL) genes expressed during G1. Cold Spring Harb Symp Quant Biol. 1991;56:69–74.
15. Motokura T, Bloom T, Kim HG, et al. A novel cyclin encoded by a *bcl1*-linked candidate oncogene. Nature. 1991;350:512–5.
16. Elledge SJ, Spottswood MR. A new human p34 protein kinase, CDK2, identified by complementation of a cdc28 mutation in *Saccharomyces cerevisiae*, is a homolog of Xenopus Eg1. EMBO J. 1991;10:2653–9.
17. Tsai LH, Harlow E, Meyerson M. Isolation of the human cdk2 gene that encodes the cyclin A- and adenovirus E1A-associated p33 kinase. Nature. 1991;353:174–7.
18. Rosenblatt J, Gu Y, Morgan DO. Human cyclin-dependent kinase 2 is activated during the S and G2 phases of the cell cycle and associates with cyclin A. Proc Natl Acad Sci U S A. 1992;89:2824–8.
19. Lew DJ, Dulić V, Reed SI. Isolation of three novel human cyclins by rescue of G1 cyclin (Cln) function in yeast. Cell. 1991;66:1197–206.
20. Koff A, Cross F, Fisher A, et al. Human cyclin E, a new cyclin that interacts with two members of the CDC2 gene family. Cell. 1991;66:1217–28.
21. Ren S, Rollins BJ. Cyclin C/cdk3 promotes Rb-dependent G0 exit. Cell. 2004;117:239–51.
22. Tassan JP, Jaquenoud M, Léopold P, et al. Identification of human cyclin-dependent kinase 8, a putative protein kinase partner for cyclin C. Proc Natl Acad Sci U S A. 1995;92:8871–5.
23. Dulić V, Lees E, Reed SI. Association of human cyclin E with a periodic G1-S phase protein kinase. Science. 1992;257:1958–61.
24. Koff A, Giordano A, Desai D, et al. Formation and activation of a cyclin E-cdk2 complex during the G1 phase of the human cell cycle. Science. 1992;257:1689–94.
25. Meyerson M, Enders GH, Wu CL, et al. A family of human cdc2-related protein kinases. EMBO J. 1992;11:2909–17.
26. Matsushime H, Ewen ME, Strom DK, et al. Identification and properties of an atypical catalytic subunit (p34$^{PSK J3}$/cdk4) for mammalian D type G1 cyclins. Cell. 1992;71:323–34.

27. Hanks SK. Homology probing: identification of cDNA clones encoding members of the protein-serine kinase family. Proc Natl Acad Sci U S A. 1987;84:388–92.
28. Weinberg RA. The retinoblastoma protein and cell cycle control. Cell. 1995;81:323–30.
29. Buchkovich K, Duffy LA, Harlow E. The retinoblastoma protein is phosphorylated during specific phases of the cell cycle. Cell. 1989;58:1097–105.
30. DeCaprio JA, Ludlow JW, Lynch D, et al. The product of the retinoblastoma susceptibility gene has properties of a cell cycle regulatory element. Cell. 1989;58:1085–95.
31. Chen PL, Scully P, Shew JY, et al. Phosphorylation of the retinoblastoma gene product is modulated during the cell cycle and cellular differentiation. Cell. 1989;58:1193–8.
32. Mihara K, Cao XR, Yen A, et al. Cell cycle-dependent regulation of phosphorylation of the human retinoblastoma gene product. Science. 1989;246:1300–3.
33. Whyte P, Buchkovich KJ, Horowitz JM, et al. Association between an oncogene and an anti-oncogene: the adenovirus E1A proteins bind to the retinoblastoma gene product. Nature. 1988;334:124–9.
34. DeCaprio JA, Ludlow JW, Figge J, et al. SV40 large tumor antigen forms a specific complex with the product of the retinoblastoma susceptibility gene. Cell. 1988;54:275–83.
35. Dyson N, Howley PM, Münger K, et al. The human papilloma virus-16 E7 oncoprotein is able to bind to the retinoblastoma gene product. Science. 1989;243:934–7.
36. Ludlow JW, DeCaprio JA, Huang CM, et al. SV40 large T antigen binds preferentially to an underphosphorylated member of the retinoblastoma susceptibility gene product family. Cell. 1989;56:57–65.
37. Lees JA, Buchkovich KJ, Marshak DR, et al. The retinoblastoma protein is phosphorylated on multiple sites by human cdc2. EMBO J. 1991;10:4279–90.
38. Mittnacht S, Hinds PW, Dowdy SF, et al. Modulation of retinoblastoma protein activity during the cell cycle. Cold Spring Harb Symp Quant Biol. 1991;56:197–209.
39. Hinds PW, Mittnacht S, Dulic V, et al. Regulation of retinoblastoma protein functions by ectopic expression of human cyclins. Cell. 1992;70:993–1006.
40. Cobrinik D, Dowdy SF, Hinds PW, et al. The retinoblastoma protein and the regulation of cell cycling. Trends Biochem Sci. 1992;17:312–5.
41. Kato J-Y, Matsushime H, Hiebert SW, et al. Direct binding of cyclin D to the retinoblastoma gene product (pRb) and pRb phosphorylation by the cyclin D-dependent kinase CDK4. Genes Dev. 1993;7:331–42.
42. Ewen ME, Sluss HK, Sherr CJ, et al. Functional interactions of the retinoblastoma protein with mammalian D-type cyclins. Cell. 1993;73:487–97.
43. Dowdy SF, Hinds PW, Louie K, et al. Physical interaction of the retinoblastoma protein with human cyclins. Cell. 1993;73:499–511.
44. Sherr CJ. Cancer cell cycles. Science. 1996;274:1672–7.
45. Matsushime H, Quelle DE, Shurtleff SA, et al. D-type cyclin-dependent kinase activity in mammalian cells. Mol Cell Biol. 1994;14:2066–76.
46. Ludlow JW, Glendening CL, Livingston DM. The retinoblastoma susceptibility gene product undergoes cell cycle-dependent dephosphorylation and binding to and release from SV40 large T. Cell. 1990;60:387–96.
47. Meyerson M, Harlow E. Identification of G1 kinase activity for cdk6, a novel cyclin D partner. Mol Cell Biol. 1994;14:2077–86.
48. Ajchenbaum F, Ando K, DeCaprio JA, et al. Independent regulation of human D-type cyclin gene expression during G1 phase in primary human T lymphocytes. J Biol Chem. 1993;268:4113–9.
49. Albanese C, Johnson J, Watanabe G, et al. Transforming p21ras mutants and c-Ets-2 activate the cyclin D1 promoter through distinguishable regions. J Biol Chem. 1995;270:23589–97.
50. Lavoie JN, L'Allemain G, Brunet A, et al. Cyclin D1 expression is regulated positively by the p42/p44MAPK and negatively by the p38/HOGMAPK pathway. J Biol Chem. 1996;271:20608–16.

51. Aktas H, Cai H, Cooper GM. Ras links growth factor signaling to the cell cycle machinery via regulation of cyclin D1 and the Cdk inhibitor p27KIP1. Mol Cell Biol. 1997;17:3850–7.
52. Weber JD, Raben DM, Phillips PJ, et al. Sustained activation of extracellular-signal-regulated kinase 1 (ERK1) is required for the continued expression of cyclin D1 in G1 phase. Biochem J. 1997;326:61–8.
53. Tetsu O, McCormick F. Beta-catenin regulates expression of cyclin D1 in colon carcinoma cells. Nature. 1999;398:422–6.
54. Rimerman RA, Gellert-Randleman A, Diehl JA. Wnt1 and MEK1 cooperate to promote cyclin D1 accumulation and cellular transformation. J Biol Chem. 2000;275:14736–42.
55. Watts CK, Sweeney KJ, Warlters A, et al. Antiestrogen regulation of cell cycle progression and cyclin D1 gene expression in MCF-7 human breast cancer cells. Breast Cancer Res Treat. 1994;31:95–105.
56. Musgrove EA, Lee CS, Buckley MF, et al. Cyclin D1 induction in breast cancer cells shortens G1 and is sufficient for cells arrested in G1 to complete the cell cycle. Proc Natl Acad Sci U S A. 1994;91:8022–6.
57. Musgrove EA, Sutherland RL. Cell cycle control by steroid hormones. Semin Cancer Biol. 1994;5:381–9.
58. Diehl JA, Zindy F, Sherr CJ. Inhibition of cyclin D1 phosphorylation on threonine-286 prevents its rapid degradation via the ubiquitin-proteasome pathway. Genes Dev. 1997;11:957–72.
59. Cheng M, Sexl V, Sherr CJ, et al. Assembly of cyclin D-dependent kinase and titration of p27Kip1 regulated by mitogen-activated protein kinase kinase (MEK1). Proc Natl Acad Sci U S A. 1998;95:1091–6.
60. Polyak K, Kato JY, Solomon MJ, et al. p27Kip1, a cyclin-Cdk inhibitor, links transforming growth factor-beta and contact inhibition to cell cycle arrest. Genes Dev. 1994;8:9–22.
61. Toyoshima H, Hunter T. p27, a novel inhibitor of G1 cyclin-Cdk protein kinase activity, is related to p21. Cell. 1994;78:67–74.
62. Soos TJ, Kiyokawa H, Yan JS, et al. Formation of p27-CDK complexes during the human mitotic cell cycle. Cell Growth Differ. 1996;7:135–46.
63. Blain SW, Montalvo E, Massagué J. Differential interaction of the cyclin-dependent kinase (Cdk) inhibitor p27Kip1 with cyclin A-Cdk2 and cyclin D2-Cdk4. J Biol Chem. 1997;272:25863–72.
64. LaBaer J, Garrett MD, Stevenson LF, et al. New functional activities for the p21 family of CDK inhibitors. Genes Dev. 1997;11:847–62.
65. Mahony D, Parry DA, Lees E. Active cdk6 complexes are predominantly nuclear and represent only a minority of the cdk6 in T cells. Oncogene. 1998;16:603–11.
66. McConnell BB, Gregory FJ, Stott FJ, et al. Induced expression of p16(INK4a) inhibits both CDK4- and CDK2-associated kinase activity by reassortment of cyclin-CDK-inhibitor complexes. Mol Cell Biol. 1999;19:1981–9.
67. Cheng M, Olivier P, Diehl JA, et al. The p21(Cip1) and p27(Kip1) CDK 'inhibitors' are essential activators of cyclin D-dependent kinases in murine fibroblasts. EMBO J. 1999;18:1571–83.
68. Geng Y, Yu Q, Sicinska E, et al. Deletion of the p27Kip1 gene restores normal development in cyclin D1-deficient mice. Proc Natl Acad Sci U S A. 2001;98:194–9.
69. Tong W, Pollard JW. Genetic evidence for the interactions of cyclin D1 and p27(Kip1) in mice. Mol Cell Biol. 2001;21:1319–28.
70. Grimmler M, Wang Y, Mund T, et al. Cdk-inhibitory activity and stability of p27Kip1 are directly regulated by oncogenic tyrosine kinases. Cell. 2007;128:269–80.
71. Larrea MD, Liang J, Da Silva T, et al. Phosphorylation of p27Kip1 regulates assembly and activation of cyclin D1-Cdk4. Mol Cell Biol. 2008;28:6462–72.
72. Jäkel H, Weinl C, Hengst L. Phosphorylation of p27Kip1 by JAK2 directly links cytokine receptor signaling to cell cycle control. Oncogene. 2011;30:3502–12.
73. Huang Y, Yoon MK, Otieno S, et al. The activity and stability of the intrinsically disordered Cip/Kip protein family are regulated by non-receptor tyrosine kinases. J Mol Biol. 2015;427:371–86.

74. Sherr CJ, Roberts JM. CDK inhibitors: positive and negative regulators of G1-phase progression. Genes Dev. 1999;13:1501–12.
75. Sheaff RM, Groudine M, Gordon J, et al. Cyclin E-CDK2 is a regulator of p27Kip1. Genes Dev. 1997;11:1464–78.
76. Vlach J, Hennecke S, Amati B. Phosphorylation-dependent degradation of the cyclin-dependent kinase inhibitor p27[Kip1]. EMBO J. 1997;16:5334–44.
77. Dyson N. The regulation of E2F by pRB-family proteins. Genes Dev. 1998;12:2245–62.
78. Nevins JR. Toward an understanding of the functional complexity of the E2F and retinoblastoma families. Cell Growth Differ. 1998;9:585–93.
79. Trimarchi JM, Lees JA. Sibling rivalry in the E2F family. Nat Rev Mol Cell Biol. 2002;3:11–20.
80. Balciunaite E, Spektor A, Lents NH, et al. Pocket protein complexes are recruited to distinct targets in quiescent and proliferating cells. Mol Cell Biol. 2005;25:8166–78.
81. Fisher RP, Morgan DO. A novel cyclin associates with MO15/CDK7 to form the CDK-activating kinase. Cell. 1994;78:713–24.
82. Matsuoka M, Kato JY, Fisher RP, et al. Activation of cyclin-dependent kinase 4 (cdk4) by mouse MO15-associated kinase. Mol Cell Biol. 1994;14:7265–75.
83. Mäkelä TP, Tassan JP, Nigg EA, et al. A cyclin associated with the CDK-activating kinase MO15. Nature. 1994;371:254–7.
84. Baldin V, Lukas J, Marcote MJ, et al. Cyclin D1 is a nuclear protein required for cell cycle progression in G1. Genes Dev. 1993;7:812–21.
85. Diehl JA, Cheng M, Roussel MF, et al. Glycogen synthase kinase-3 beta regulates cyclin D1 proteolysis and subcellular localization. Genes Dev. 1998;12:3499–511.
86. Alt JR, Cleveland JL, Hannink M, et al. Phosphorylation-dependent regulation of cyclin D1 nuclear export and cyclin D1-dependent cellular transformation. Genes Dev. 2000;14:3102–14.
87. Fantl V, Stamp G, Andrews A, et al. Mice lacking cyclin D1 are small and show defects in eye and mammary gland development. Genes Dev. 1995;9:2364–72.
88. Sicinski P, Donaher JL, Parker SB, et al. Cyclin D1 provides a link between development and oncogenesis in the retina and breast. Cell. 1995;82:621–30.
89. Yu Q, Geng Y, Sicinski P. Specific protection against breast cancers by cyclin D1 deletion. Nature. 2001;411:1017–21.
90. Ma C, Papermaster D, Cepko CL. A unique pattern of photoreceptor degeneration in cyclin D1 mutant mice. Proc Natl Acad Sci U S A. 1998;95:9938–43.
91. Sicinski P, Donaher JL, Geng Y, et al. Cyclin D2 is an FSH-responsive gene involved in gonadal cell proliferation and oncogenesis. Nature. 1996;384:470–4.
92. Huard JM, Forster CC, Carter ML, et al. Cerebellar histogenesis is disturbed in mice lacking cyclin D2. Development. 1999;126:1927–35.
93. Lam EW, Glassford J, Banerji L, et al. Cyclin D3 compensates for loss of cyclin D2 in mouse B-lymphocytes activated via the antigen receptor and CD40. J Biol Chem. 2000;275:3479–84.
94. Solvason N, Wu WW, Parry D. Cyclin D2 is essential for BCR-mediated proliferation and CD5 B cell development. Int Immunol. 2000;12:631–8.
95. Kushner JA, Ciemerych MA, Sicinska E, et al. Cyclins D2 and D1 are essential for postnatal pancreatic beta-cell growth. Mol Cell Biol. 2005;25:3752–62.
96. Kowalczyk A, Filipkowski RK, Rylski M. The critical role of cyclin D2 in adult neurogenesis. J Cell Biol. 2004;167:209–13.
97. Sicinska E, Aifantis I, Le Cam L, et al. Requirement for cyclin D3 in lymphocyte development and T cell leukemias. Cancer Cell. 2003;4:451–61.
98. Cooper AB, Sawai CM, Sicinska E, et al. A unique function for cyclin D3 in early B cell development. Nat Immunol. 2006;7:489–97.
99. Sicinska E, Lee YM, Gits J, et al. Essential role for cyclin D3 in granulocyte colony-stimulating factor-driven expansion of neutrophil granulocytes. Mol Cell Biol. 2006;26:8052–60.
100. Peled JU, Yu JJ, Venkatesh J, et al. Requirement for cyclin D3 in germinal center formation and function. Cell Res. 2010;20:631–46.

101. Ciemerych MA, Kenney AM, Sicinska E, et al. Development of mice expressing a single D-type cyclin. Genes Dev. 2002;16:3277–89.
102. Kozar K, Ciemerych MA, Rebel VI, et al. Mouse development and cell proliferation in the absence of D-cyclins. Cell. 2004;118:477–91.
103. Carthon BC, Neumann CA, Das M, et al. Genetic replacement of cyclin D1 function in mouse development by cyclin D2. Mol Cell Biol. 2005;25:1081–8.
104. Sawai CM, Freund J, Oh P, et al. Therapeutic targeting of the cyclin D3:CDK4/6 complex in T cell leukemia. Cancer Cell. 2012;22:452–65.
105. Rane SG, Dubus P, Mettus RV, et al. Loss of Cdk4 expression causes insulin-deficient diabetes and Cdk4 activation results in beta-islet cell hyperplasia. Nat Genet. 1999;22:44–52.
106. Tsutsui T, Hesabi B, Moons DS, et al. Targeted disruption of CDK4 delays cell cycle entry with enhanced p27(Kip1) activity. Mol Cell Biol. 1999;19:7011–9.
107. Malumbres M, Sotillo R, Santamaría D, et al. Mammalian cells cycle without the D-type cyclin-dependent kinases Cdk4 and Cdk6. Cell. 2004;118:493–504.
108. Martín J, Hunt SL, Dubus P, et al. Genetic rescue of Cdk4 null mice restores pancreatic beta-cell proliferation but not homeostatic cell number. Oncogene. 2003;22:5261–9.
109. Sankaran VG, Ludwig LS, Sicinska E, et al. Cyclin D3 coordinates the cell cycle during differentiation to regulate erythrocyte size and number. Genes Dev. 2012;26:2075–87.
110. Williams ME, Swerdlow SH, Meeker TC. Chromosome t(11;14)(q13;q32) breakpoints in centrocytic lymphoma are highly localized at the bcl-1 major translocation cluster. Leukemia. 1993;7:1437–40.
111. Williams ME, Swerdlow SH. Cyclin D1 overexpression in non-Hodgkin's lymphoma with chromosome 11 bcl-1 rearrangement. Ann Oncol. 1994;Suppl 1:71–3.
112. Salaverria I, Royo C, Carvajal-Cuenca A, et al. CCND2 rearrangements are the most frequent genetic events in cyclin D1(−) mantle cell lymphoma. Blood. 2013;121:1394–402.
113. Hall M, Peters G. Genetic alterations of cyclins, cyclin-dependent kinases, and Cdk inhibitors in human cancer. Adv Cancer Res. 1996;68:67–108.
114. Beroukhim R, Mermel CH, Porter D, et al. The landscape of somatic copy-number alteration across human cancers. Nature. 2010;463:899–905.
115. Kim JK, Diehl JA. Nuclear cyclin D1: an oncogenic driver in human cancer. J Cell Physiol. 2009;220:292–6.
116. Delmer A, Ajchenbaum-Cymbalista F, Tang R, et al. Overexpression of cyclin D2 in chronic B-cell malignancies. Blood. 1995;85:2870–6.
117. Suzuki R, Kuroda H, Komatsu H, et al. Selective usage of D-type cyclins in lymphoid malignancies. Leukemia. 1999;13:1335–42.
118. Sonoki T, Harder L, Horsman DE, et al. Cyclin D3 is a target gene of t(6; 14)(p21.1;q32.3) of mature B-cell malignancies. Blood. 2001;98:2837–44.
119. Clappier E, Cuccuini W, Cayuela JM, et al. Cyclin D2 dysregulation by chromosomal translocations to TCR loci in T-cell acute lymphoblastic leukemias. Leukemia. 2006;20:82–6.
120. Igawa T, Sato Y, Takata K. Cyclin D2 is overexpressed in proliferation centers of chronic lymphocytic leukemia/small lymphocytic lymphoma. Cancer Sci. 2011;102:2103–7.
121. Schmitz R, Young RM, Ceribelli M, et al. Burkitt lymphoma pathogenesis and therapeutic targets from structural and functional genomics. Nature. 2012;490:116–20.
122. Li ZM, Spagnuolo L, Mensah AA, et al. Gains of CCND3 gene in ocular adnexal MALT lymphomas: an integrated analysis. Br J Haematol. 2013;160:719–22.
123. Khatib ZA, Matsushime H, Valentine M, et al. Coamplification of the CDK4 gene with MDM2 and GLI in human sarcomas. Cancer Res. 1993;53:5535–41.
124. He J, Allen JR, Collins VP, et al. CDK4 amplification is an alternative mechanism to p16 gene homozygous deletion in glioma cell lines. Cancer Res. 1994;54:5804–7.
125. Reifenberger G, Reifenberger J, Ichimura K, et al. Amplification of multiple genes from chromosomal region 12q13-14 in human malignant gliomas: preliminary mapping of the amplicons shows preferential involvement of CDK4, SAS, and MDM2. Cancer Res. 1994;54:4299–303.

126. Schmidt EE, Ichimura K, Reifenberger G, et al. CDKN2 (p16/MTS1) gene deletion or CDK4 amplification occurs in the majority of glioblastomas. Cancer Res. 1994;54:6321–4.
127. Costello JF, Plass C, Arap W, et al. Cyclin-dependent kinase 6 (CDK6) amplification in human gliomas identified using two-dimensional separation of genomic DNA. Cancer Res. 1997;57:1250–4.
128. Nagel S, Leich E, Quentmeier H, et al. Amplification at 7q22 targets cyclin-dependent kinase 6 in T-cell lymphoma. Leukemia. 2007;22:387–92.
129. van Dekken H, van Marion R, Vissers KJ, et al. Molecular dissection of the chromosome band 7q21 amplicon in gastroesophageal junction adenocarcinomas identifies cyclin-dependent kinase 6 at both genomic and protein expression levels. Genes Chromosomes Cancer. 2008;47:649–56.
130. Tsai JW, Li CF, Kao YC, et al. Recurrent amplification at 7q21.2 targets CDK6 gene in primary myxofibrosarcomas and identifies CDK6 overexpression as an independent adverse prognosticator. Ann Surg Oncol. 2012;19:2716–25.
131. Serrano M, Hannon GJ, Beach D. A new regulatory motif in cell-cycle control causing specific inhibition of cyclin D/CDK4. Nature. 1993;366:704–7.
132. Lukas J, Parry D, Aagaard L, et al. Retinoblastoma-protein-dependent cell-cycle inhibition by the tumour suppressor p16. Nature. 1995;375:503–6.
133. Koh J, Enders GH, Dynlacht BD, et al. Tumour-derived p16 alleles encoding proteins defective in cell-cycle inhibition. Nature. 1995;375:506–10.
134. Medema RH, Herrera RE, Lam F, et al. Growth suppression by p16ink4 requires functional retinoblastoma protein. Proc Natl Acad Sci U S A. 1995;92:6289–9623.
135. Kamb A, Gruis NA, Weaver-Feldhaus J, et al. A cell cycle regulator potentially involved in genesis of many tumor types. Science. 1994;264:436–40.
136. Hannon GJ, Beach D. p15INK4B is a potential effector of TGF-beta-induced cell cycle arrest. Nature. 1994;371:257–61.
137. Quelle DE, Zindy F, Ashmun RA, et al. Alternative reading frames of the INK4a tumor suppressor gene encode two unrelated proteins capable of cell cycle arrest. Cell. 1995;83:993–1000.
138. Kamijo T, Zindy F, Roussel MF, et al. Tumor suppression at the mouse INK4a locus mediated by the alternative reading frame product p19ARF. Cell. 1997;91:639–59.
139. Ranade K, Hussussian CJ, Sikorski RS, et al. Mutations associated with familial melanoma impair p16INK4 function. Nat Genet. 1995;10:114–6.
140. Bartek J, Bartkova J, Lukas J. The retinoblastoma protein pathway and the restriction point. Curr Opin Cell Biol. 1996;8:805–14.
141. Finn RS, Crown JP, Lang I, et al. The cyclin-dependent kinase 4/6 inhibitor palbociclib in combination with letrozole versus letrozole alone as first-line treatment of oestrogen receptor-positive, HER2-negative, advanced breast cancer (PALOMA-1/TRIO-18): a randomised phase 2 study. Lancet Oncol. 2015;16:25–35.
142. Krishnamurthy J, Ramsey MR, Ligon KL, et al. p16INK4a induces an age-dependent decline in islet regenerative potential. Nature. 2006;443:453–7.
143. Musgrove EA, Caldon CE, Barraclough J, et al. Cyclin D as a therapeutic target in cancer. Nat Rev Cancer. 2011;11:558–72.
144. Reddy HK, Mettus RV, Rane SG, et al. Cyclin-dependent kinase 4 expression is essential for neu-induced breast tumorigenesis. Cancer Res. 2005;65:10174–8.
145. Yu Q, Sicinska E, Geng Y, et al. Requirement for CDK4 kinase function in breast cancer. Cancer Cell. 2006;9:23–32.
146. Landis MW, Pawlyk BS, Li T, et al. Cyclin D1-dependent kinase activity in murine development and mammary tumorigenesis. Cancer Cell. 2006;9:13–22.
147. Choi YJ, Li X, Hydbring P, et al. The requirement for cyclin D function in tumor maintenance. Cancer Cell. 2012;22:438–51.
148. Garber K. The cancer drug that almost wasn't. Science. 2014;345:865–7.
149. Sherr CJ, Beach D, Shapiro GI. Targeting CDK4 and CDK6: from discovery to therapy. Cancer Discov. 2016;6:353–67.

Chapter 2
Mammalian Development and Cancer: A Brief History of Mice Lacking D-Type Cyclins or CDK4/CDK6

Ilona Kalaszczynska and Maria A. Ciemerych

Abstract Cellular proliferation is controlled by the orchestrated action of many cell cycle regulators. Among them are cyclins and cyclin-dependent kinases (CDKs), the activity of which is necessary to drive each phase of the cell cycle. Mitogenic stimulation leads to the expression of D cyclins which bind and activate CDK4 and CDK6. This event triggers a chain of events ultimately leading to cell division. Dissection of the functions of D cyclins and CDK4 in the regulation of proliferation of mammalian cells was greatly facilitated by the generation of genetically modified mice in which either D cyclins, their CDK partners, or other cell cycle regulators were ablated or replaced. In general, variable impact of germline loss of these cell cycle regulators on different tissues underscores specific roles for D cyclins and their partner CDKs in differentiation and development. These mouse models have also proved crucial for studies analyzing tumor development and for the discovery and evaluation of anticancer therapies, often linking tissue-specific functions to antineoplastic effects of inhibition of cyclin-D-dependent processes. This chapter summarizes the history of mice lacking D cyclins and CDK4/CDK6 and presents a synopsis of key findings from those animal models.

Keywords Mammals • Mice • Knock-out mice • Knock-in mice • Cell cycle • Cyclin D • CDK4 • CDK6 • Embryogenesis • Cancer

I. Kalaszczynska
Department of Histology and Embryology, Center for Biostructure Research,
Medical University of Warsaw, Warsaw, Poland

Centre for Preclinical Research and Technology, Warsaw, Poland

M.A. Ciemerych (✉)
Department of Cytology, Institute of Zoology, Faculty of Biology,
University of Warsaw, Ilji Miecznikowa 1, 02-096 Warsaw, Poland
e-mail: ciemerych@biol.uw.edu.pl

© Springer International Publishing AG 2018
P.W. Hinds, N.E. Brown (eds.), *D-type Cyclins and Cancer*,
Current Cancer Research, DOI 10.1007/978-3-319-64451-6_2

2.1   Introduction

In 1855 Rudolf Virchow formulated his famous *Omnis cellula e cellula*—all cells come from cells. He also explained that "we must reduce all tissues to a single simple element, the cell (...), and from it emanate all the activities of life both in health and in sickness" [216]. Thus, to understand how the organism originates, develops, and matures, we have to understand how its cells proliferate, differentiate, become quiescent, die, or transform to malfunction and cause disease, such as cancer. Importantly, unicellular organisms, as well as cells of multicellular organisms, exploit the same molecular machinery governing their proliferation. This machinery ensures that the newly formed cell becomes ready to replicate its genetic material and that any mistakes occurring during replication will be removed. Next, this process dictates that cell division will produce two daughter cells properly prepared for either the next cell cycle or another fate, such as differentiation. Due to the wide variety of dividing cells, some aspects of cell cycle progression may be modified; however, the core of this process is constant and relies on the function of cyclin-dependent kinases (CDKs) and their regulatory cofactors (cyclins).

In this chapter we will present one of the crucial cell cycle regulators, D-type cyclins and their CDK partners. We will center on the "ab ovo" part of the characterization of these proteins, i.e., their role in the development, and also outline their involvement in carcinogenesis. The majority of the studies presented here would not have been possible without the groundbreaking discoveries and techniques developed by Martin Evans and Matthew Kaufman as well as Gail Martin, who derived the first lines of mouse embryonic stem cells [58, 146], and Mario Capecchi and Oliver Smithies who showed how to genetically modify these cells [141]. Important input also came from Andrzej K. Tarkowski who was the first to create chimeric mice [229, 230], which established the basis for an indispensable method to generate knock-out or knock-in mice. Our major goal is to summarize what has been learned using mice in which cyclins D, CDK4, CDK6, or other cell cycle regulators were ablated or replaced. We are aware, however, that presenting all of the currently available data is not possible. Thus, we do regret the omission of any relevant finding and view. We are sure, however, that other chapters presented in this book will expand upon our summary of progress made in understanding D-type cyclin function derived from genetically modified mice.

2.2   The Core

The first studies leading to the discovery of the universal mechanisms governing cell cycle progression focused on *Rana pipiens* oocytes undergoing meiotic division—the so-called meiotic maturation. Experiments demonstrating that cytoplasm from dividing frog oocyte induced meiosis in prophase oocytes led to the discovery of the activity described as MPF (maturation-promoting factor or M-phase-promoting factor [148, 149, 219]). In a short time, similar activity was confirmed in mouse oocytes [11] and also in dividing somatic cells [183, 221] proving that MPF triggers not only meiosis but also mitosis and is responsible for interphase/metaphase transition. Next, a drop in MPF activity was shown to be prerequisite for metaphase exit. The biochemical nature of MPF was soon revealed—it was characterized as a complex of a protein kinase, later termed cyclin-dependent kinase (CDK), and a regulatory component, cyclin. The first CDK, i.e., CDK1, was discovered in yeast by cloning *cdc2* and *CDC28* [81, 82, 125, 215]. CDK1 activators were identified during analyses of dividing sea urchin and clam embryos [59, 192, 224]. They were named cyclins due to their periodic expression pattern. Cyclins accumulated in interphase and were abruptly degraded in M-phase just before each cleavage division of an embryo. These milestone discoveries were soon followed by characterization of other cyclins, CDKs, and their positive and negative regulators, present not only in yeast and animal cells but also in plant cells. It was also shown that specific CDKs can be regulated only by particular cyclins, the synthesis of which leads to activation of these enzymes. Precisely orchestrated destruction of the cyclins results in a drop in CDK activity.

Next, G1- and S-phase-specific CDK-cyclin complexes were identified along with their cell cycle-specific substrates. Thus, cyclin D-CDK4 or CDK6 (CDK4/CDK6) complexes regulate G1 phase, CDK2 together with E- and A-type cyclins controls S phase, and CDK1 activated by A- and B-type cyclins coordinates M-phase progression (Fig. 2.1). Current evidence supports a simplified model

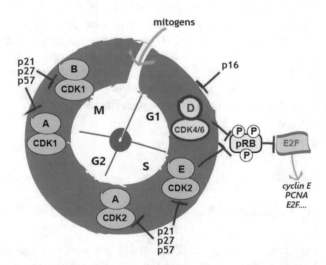

Fig. 2.1 Simplified summary of cell cycle regulation. Cell cycle progression is precisely controlled by periodically active cyclin-CDK complexes. Mitogen stimulation drives cell cycle reentry by induction of D-type cyclin synthesis. Once synthesized, D cyclins bind and activate CDK4 and CDK6 kinase subunits, which in turn phosphorylate and inactivate pRb allowing activation of E2F transcription factors and their cofactors. One of the first products of E2F-controlled transcription is cyclin E which binds and activates CDK2 allowing initiation of S-phase. Next, S-phase is controlled by cyclin A-CDK2 complexes. Cyclin A-CDK1 activation that is necessary for M-phase entry is followed by cyclin B-CDK1 activation. Completion of each cell cycle phase requires degradation of the specific cyclin and, as a result, CDK inactivation. Each of the CDKs is blocked by specific inhibitors—CDK4/CDK6 by INK4 family members, e.g., p16[Ink4a], CDK2, and CDK1 by KIP/CIP inhibitors, i.e., p21[cip1], p27[kip1], and p57[kip2] (Modified from Ciemerych et al. [42])

showing that in order to be active, a CDK has to be postranslationally modified. One of the crucial modifications is introduced by CAK, i.e., CDK-activating kinase that phosphorylates the T-loop of monomeric CDK [66]. Interestingly, CAK is itself composed of a CDK, CDK7, which is activated by cyclin H and MAT1. Further, in addition to its involvement in cell cycle control, CDK7 is also a component of transcription factor TFIIH, which plays a role in the regulation of gene expression [34] as do other CDKs, such as CDK8 or CDK9 (for review see [142]).

In addition to activation by posttranslational modification, CDKs are also subject to inhibitory phosphorylation, e.g., p-T14/Y15 in case of CDK1, which needs to be removed by CDK-specific phosphatases [208], and also to inhibition by members of the INK family (e.g., p16^{Ink4a}—specific for CDK4 and CDK6) or CIP/KIP family (e.g., p21^{Cip1}, p27^{Kip1}, p57^{Kip2}—specific for CDK2 and CDK1) of protein inhibitors. Interestingly, cyclin D-CDK4/CDK6 complexes bind and sequester CIP/KIP proteins, such as p27^{kip1}, avoiding being inhibited by them and promoting activation of other CDKs [9, 23, 114]. Nevertheless, it is the binding of CDK and cyclin that is a sine qua non condition for CDK activation, with each of the other events described serving to add exquisite layers of regulatory control on these vital cell cycle control enzymes.

The core of the cell cycle regulatory machinery is operative in dividing cells. However, one has to be aware that many cell types utilize customized adjustments to fundamental cell cycle mechanisms. These sometimes subtle differences allow certain cells to adapt to specific developmental or environmental requirements. For example, proliferation of certain embryonic cells, such as embryonic stem cells (ESCs), is not inhibited by p16^{Ink4a} [201] raising the possibility that the cyclin D-CDK4/CDK6 pathway might be modified or not fully operative during early mammalian development [61, 238]. Thus, the regulation of cleavage divisions in developing embryos and proliferation of precursors of various cell types, or embryonic stem cells, can serve as illustration of such fine-tuning (for review see, e.g., [42, 76, 111, 161, 202]).

2.3   The Details

2.3.1   D-Type Cyclins: What Are They and What Do They Do?

Three D-type cyclins, i.e., cyclin D1, D2, and D3, are present in mammalian cells and tissues. They were first reported as products of genes responding to mitogen stimulation and involved in G1 phase regulation [128, 153, 165, 166, 241, 244, 245]. They are encoded by separate genes but share significant amino acid identity that reaches 50–60% throughout the entire coding sequence and 75–78% within the most conserved cyclin box domain [91, 245]. As was mentioned above, expression of D-type cyclins is largely controlled by the extracellular environment—they are upregulated during cell cycle entry as a result of signals coming from extracellular matrix or soluble mitogens reaching the cell [106]. For example, cyclin D1 levels can be increased by mitogenic stimulation activating the MAPK canonical pathway, i.e., Ras-Raf-MEK-ERK1/ERK2 [3, 120], PI3K, Wnt, or other signaling pathways [106, 167]. Conversely, D-type cyclin expression declines when anti-mitogens are added and for these reasons they might be described as sensors of environmental changes.

The first identified function of D-type cyclins was to control cell cycle reentry by activating CDK4/CDK6 which phosphorylates pRb family members, i.e., pRb, p107, and p130 [18, 154, 155, 158, 253]. In its active, i.e., hypophosphorylated, state, pRb binds and prevents activity of the E2F transcription factors. Phosphorylation of pRB leads to the release of E2Fs and results in the activation of E2F-controlled genes, among them those encoding E- and A-type cyclins, i.e., cyclins involved in the activation of CDK2-regulating initiation and progression of S-phase [45, 54, 209, 210]. Cyclin D-CDK4/CDK6 complexes impact cell cycle progression also by controlling other proteins, such as SMAD family members. Phosphorylation of SMAD3, a factor playing a crucial role in the antiproliferative TGF-β pathway, leads to the inhibition of its antiproliferative function [133, 156] and promotion of cell cycle progression. Systemic screening for cyclin D1-CDK4/CDK6 and cyclin D3-CDK4/CDK6 substrates revealed, apart from pRB family members and SMAD3, another 68 potential targets for these kinases [4]. Among identified targets were such factors as Myc or forkhead box M1 (FOXM1), proteins which when phosphorylated and stabilized activate the expression of G1/S phase genes [4]. Interestingly, the number of substrates uncovered as a result of these analyses seemed to depend on the cyclin D type. Cyclin D1-CDK4/CDK6 substrates were less abundant than those phosphorylated by cyclin D3-CDK4/CDK6. In recent years CDK-independent functions of D-type cyclins were also uncovered (see below).

2.3.2   D-Type Cyclins: Where Are They Expressed?

All three D-type cyclins, as well as their CDK partners, are detectable during oogenesis [107, 164], spermatogenesis [19, 103, 254], and also at each step of pre- and post-implantation mammalian development [40]. Interestingly, they are expressed with significant overlap (Fig. 2.2). For example, in the developing nervous system, cyclins D1 and D2 are detectable in distinct cellular compartments, and their synthesis dynamically changes along the course of development [2, 239]. In some tissues, such as stratified squamous epithelia, cyclin D1 synthesis is associated with proliferating cells, whereas cyclin D3 is present in cells more advanced in differentiation [15]. In proliferating skeletal myoblasts, cyclin D1 prevents exit from the cell cycle and terminal differentiation [184, 217, 218]. Thus, the formation of mature myotubes is associated with decrease in cyclins D1 and D2 and increase in cyclin D3 expression [31, 104, 145]. In embryonic and also in adult tissues, some cellular compartments express a combination of two or even three D-type cyclins (e.g., [15, 16, 41, 73, 175, 186, 226, 248]) (Fig. 2.2). Expression of CDK4 and CDK6 does not seem to be so finely assigned as is the case for D-type cyclins [41].

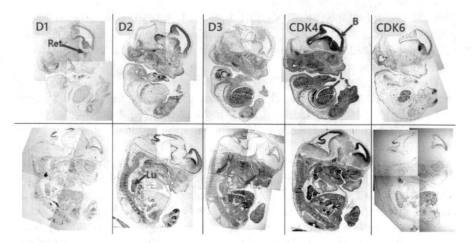

Fig. 2.2 Cyclin D, CDK4, and CDK6 expression in E13.5 mouse embryos. Sagittal sections of mouse embryos were hybridized with riboprobes specific for cyclin D1, cyclin D2, cyclin D3, CDK4, and CDK6 (Technical details in Ciemerych et al. [41]). *Black color* represents hybridization signal. *B* brain, *Ret* retina, *L* liver, *V vertebrae*, *H* heart, *Lu* lungs

The precisely timed expression patterns of D-type cyclins suggested that each of them may play some non-redundant and/or CDK-unrelated functions. Experiments aiming at the verification of this hypothesis started with the generation of mutant mice lacking a single D-type cyclin [62, 211, 213, 214] and soon was followed by experiments analyzing results of ablation of two and finally all three D-type cyclins [43, 109]. Next, kinase-dependent functions were tested in CDK4- and CDK6-null mice [144, 157, 182, 231]. As a result it was uncovered that lack of D-type cyclins or CDK4/CDK6 had dramatic consequences for the proper development of certain cellular compartments.

2.3.3   D-Type Cyclins and Their Partners: How to Live Without Them?

2.3.3.1   Single Knock-Out Mice

In 1995, the phenotype of the first cyclin D knock-out mice was described. It was only a few years after D-cyclin-encoding genes were cloned, and as not much was known about their specific function, a lot was to be discovered. Although all three D cyclins showed very high sequence similarity, it was suspected that each of them could play unique functions. This notion was supported by observations showing that despite widespread expression of each D-type cyclin, phenotypes of knock-out mice were limited to a narrow subset of cellular compartments. Importantly, mice lacking CDK4 or CDK6 displayed abnormalities within tissues and organs similar to those affected by the lack of D-type cyclins suggesting that at least some phenotypes resulting from D-cyclin loss are CDK related (Table 2.1).

Table 2.1 Phenotypes of mice lacking cyclins D or CDK4/CDK6

Gene/genes disrupted	Survival	Phenotypes	References
		"Conventional" knock-outs	
Cyclin D1	Viable	Reduced body size, neurological abnormalities, hypoplastic retinas, impaired proliferation of mammary gland epithelium during pregnancy Resistant to Ras- and ErbB2-driven breast cancers, skin papillomas, Apc^min-driven intestinal polyps	[62, 89, 190, 213, 249]
Cyclin D2	Viable	Infertile females (inability of ovarian granulosa cells to proliferate in response to FSH; oocyte development not affected), males fertile but have small testes and decreased sperm counts, impaired cerebellar development, impaired proliferation of B-lymphocytes Reduced susceptibility to gonadal tumors, BCR/ABL-driven transformation of hematopoietic cells, Apc^min-driven intestinal polyps	[29, 47, 87, 95, 115, 214, 220]
Cyclin D3	Viable	Hypoplastic thymi (reduced expansion of immature T lymphocytes) Resistance to Notch-driven leukemias (T-ALL); delayed development of p56lck-driven thymomas	[211]
Cyclins D1D2	Viable but die within 3 weeks after birth	Reduced body size, hypoplastic cerebella	[43]
Cyclins D1D3	Most die immediately after birth; a fraction can survive up to 2 months	Neonatal lethality due to aspiration with meconium (probably caused by neurological defects)	[43]
Cyclins D2D3	Embryonic lethal (E16.5–17.5)	Severe megaloblastic anemia	[43]
Cyclins D1D2D3	Embryonic lethal (E16)	Severe megaloblastic anemia, multilineage hematopoietic failure, abnormal heart development	[109]
CDK4	Viable	Reduced body size, neurological abnormalities, pancreatic dysfunction resulting in diabetes (β-cell hypoplasia), males infertile due to reduced number of spermatids and mature spermatozoa, females infertile due to failure in formation of corpus luteum. Decreased incidence of skin tumors, Myc-driven tumors of oral mucosa	[157, 160, 182, 191, 231]

(continued)

Table 2.1 (continued)

Gene/genes disrupted	Survival	Phenotypes	References
CDK6	Viable	Reduced thymus and spleen cellularity, defects in expansion of immature T lymphocytes, deficiency in hematopoietic stem cell function Resistant to Akt-driven lymphomas and BCR-ABL^{p2101}-induced leukemia	[86, 144, 203]
CDK4, CDK6	Embryonic lethal (E14.5-E18.5)	Severe anemia, defects in maturation of different hematopoietic lineages	[144]
CDK4, CDK6, CDK2		Die at E13.5-15.5. Multilineage hematopoietic failure, abnormal heart development	[200]
		"Conventional"/conditional* knock-outs *gene disrupted using Cre-LoxP technique	
D1*		Loss of cyclin D1 in hepatocytes causes increased gluconeogenesis and hyperglycemia Loss of cyclin D1 in mammary glands prevents tumors induced by ErbB2 oncogene	[37, 127]
D3*		Hypoplastic thymi (reduced expansion of immature T lymphocytes) Resistance to Notch-driven leukemias (T-ALL)	[37]
D1*D2D3*		Loss of hematopoietic stem cells	[38]
D1D2D3E1*E2		Loss of D- and E-type cyclins in ES cells does not prevent proliferation Loss of D- and E-type cyclins in MEFs abolishes proliferation	[134]
CDK4* and CDK2*		No obvious abnormalities	[12]

The first cyclin studied using a genetic mouse model was cyclin D1. Its expression was independently disrupted by Sicinski (Weinberg group) and Fantl (Dickson group) [62, 213]. D1-null mice display reduced body mass, a spastic leg-clasping reflex, and a partially penetrant premature mortality within the first weeks of life. The latter phenotype is explained by abnormalities in the development and function of the nervous system. Despite the neurological deficiencies, brain size and neural progenitor cell number is comparable to that observed in wild-type controls [36, 74]. The loss of cyclin D1 does not reduce the number of neural progenitor cells in the subgranular zone (SGZ) [108]. It impacts, however, Schwann and glial cell proliferation associated with postnatal injury [7, 171] but does not markedly prevent axonal regrowth during induced regeneration [105]. Two initial studies on D1$^{-/-}$ mice revealed abnormal development of retinas [62, 213] resulting from restricted proliferation of retinal cells and increased photoreceptor cell death [137].

Interestingly, functional redundancy among D-cyclin subtypes was documented by the analysis of knock-in mice expressing cyclin D2 in place of cyclin D1 in that development of retinas was nearly normal [30]. Cyclin D2 could also replace cyclin D1 function in estrogen-induced proliferation of other tissues, i.e., mouse uterine epithelium [33]. However, other studies suggested that neither cyclin D2 nor cyclin D3 could fully ameliorate the retinal phenotype [52]. Interestingly, the function of cyclin D1 was replaceable by a downstream cell cycle regulator—cyclin E [70]. A second dramatic phenotype of D1-null mice is associated with the failure of mammary glands to undergo normal lobuloalveolar development during pregnancy [62–64, 213]. As a result $D1^{-/-}$ females cannot feed their pups. This phenotype could also be rescued by cyclin D2 [30].

Dissection of specific functions of cyclin D1 led to the generation of two knock-in mouse strains. At first, knock-in mice carrying a version of cyclin D1 that lacks the ability to activate CDK4/CDK6 were analyzed [117]. Such animals manifest only slightly underdeveloped retinas. Also pregnancy-induced mammary gland epithelial expansion is not substantially affected in uniparous knock-in females [117]. Further studies showed, however, that abrogation of cyclin D1-associated kinase activity influences mammary gland progenitor cell self-renewal and impacts their differentiation and tissue regeneration [96], as well as leads to upregulation of autophagy [27]. Similar to cyclin D1 knock-out animals, "knock-ins" are also characterized by some growth deficiency and neurological phenotypes, i.e., leg clasping. In contrast, mutation in the LxCxE motif, which is required for binding of D-type cyclins with pRb [55] and is essential for cell cycle regulatory functions of these proteins, impacts neither retinal development nor mammary gland function [10, 118]. In vitro studies exposed, however, that the LxCxE motif is crucial for cyclin D2 function, documenting that these two cyclins might not play redundant roles in cell cycle control [10].

The phenotype of cyclin D2-deficient mice is also very narrow. Females are sterile as a result of the inability of ovarian granulosa cells to proliferate in response to follicle-stimulating hormone (FSH). As a consequence, ovarian follicles do not form properly, and oocytes cannot be ovulated. $D2^{-/-}$ males are fertile but testes are hypoplastic [214]. Lack of cyclin D2 also impacts the proliferation of peripheral B-lymphocytes [115, 220] and pancreatic β-cells [71, 112, 113]. Next, several neurological phenotypes are characteristic for cyclin $D2^{-/-}$ mice. Among them are mild cerebellar abnormalities [75, 87], a decrease in intermediate progenitor cells in the embryonic cortex [74], as well as impaired adult neurogenesis [5, 108]. D3-deficient mice, in turn, are viable but display abnormalities in T and B cell [48, 150, 211] as well as erythrocyte development [199]. D3-null mice are also characterized by deficient maturation of granulocytes in bone marrow and a reduced number of granulocytes and neutrophils in the blood [212]. Interestingly, cyclin D3 was also shown to be involved in pancreatic β-cell function, since in the nonobese diabetic (NOD) type 1 diabetes-prone mouse, lack of this cyclin exacerbates diabetes and impairs glucose responsiveness [195].

D-type cyclins bind and activate CDK4 and CDK6. As with their cyclin partners, the expression of these catalytic subunits during mammalian development is generally overlapping [41] (Fig. 2.2). Both kinases were shown to share 71% amino acid identity and phosphorylate the same substrates, e.g., pRb family members; thus, it was initially widely accepted that they play a redundant function [158, 159]. However, some lines of evidence document that subtle functional differences between these two kinases might exist. For example, CDK4 was shown to preferentially phosphorylate pRb at the threonine 826 residue, while CDK6 phosphorylates threonine 821 [225]. Next, in T lymphocytes, CDK6 is activated before CDK4 [136], and their actions seem to be different in such distinct cells as thymocytes [85, 86], osteoblasts [56, 57], or astrocytes [173], strongly suggesting a tissue-specific role for CDK6 in cellular differentiation [77, 78].

Generation of CDK4- and CDK6-deficient mice showed that both mutants are viable and characterized by rather mild phenotypes, suggesting that either CDK4 and CDK6 could substitute for each other in many tissues, or they can be replaced by CDK2, which indeed is able to provide functional compensation by interacting with D-type cyclins [144]. Importantly, some defects of CDK4- and CDK6-deficient mice mimic those observed in single D-type cyclin mutants (Table 2.1). CDK4$^{-/-}$ mouse phenotypes essentially equate with those of both cyclin D1 and cyclin D2 knock-out mice, i.e., retarded growth (similar to cyclin D1$^{-/-}$), ovarian and testicular defects, and also pancreatic hypoplasia (similar to cyclin D2$^{-/-}$) [147, 157, 182, 231]. Female infertility, however, is not caused by a defect in granulosa cell proliferation, as was shown for cyclin D2$^{-/-}$ females. Rather, infertility in CDK4$^{-/-}$ mice results from a failure in the development of pituitary lactotroph cells that leads to a deficiency in prolactin production and defective formation of corpus luteum and as a consequence prevents embryo implantation [98, 162, 163, 182]. Other affected processes include adipogenesis [1] and T lymphocyte maturation [39]. Lack of CDK4 also causes some neurological deficiencies, e.g., compromised locomotion [182] and a decrease in the proliferation of Schwann cells, but only during early postnatal development [8]. CDK6 deficiency results in hematopoietic defects (similar to cyclin D3$^{-/-}$ mice) manifested by abnormal spleen and thymus development, decreased number of peripheral blood cells [86, 144], as well as a partial deficiency in hematopoietic stem cell function, i.e., impaired repopulation after competitive transplantation [203]. The fact that some cell types fail to properly develop and function when either a single D-type cyclin CDK or CDK6 is absent in tissues that express most or all of these subunits underscores their unique functions.

2.3.3.2   Double and Triple Knock-Out Mice

In 2002 we stated that "These single-knock-out experiments are illuminating, but their analyses are greatly confounded by the presence of two remaining, intact D-cyclins, which may compensate for the ablated protein. We decided to reduce this complexity by creating mouse strains expressing only a single D-type cyclin. In doing so, we hoped to be able to directly test which proliferative and developmental functions can be executed solely by cyclin D1, D2, or D3" [43]. The double knock-out mice, i.e., D1$^{-/-}$D2$^{-/-}$ (expressing only cyclin D3), D1$^{-/-}$D3$^{-/-}$ (expressing only cyclin D2), and D2$^{-/-}$D3$^{-/-}$ (expressing only cyclin D1), so-called "single-cyclin" mice, displayed the additive defects characteristic for mice lacking a single D-type cyclin. Animals expressing only cyclin D3 were born alive but died within 3 weeks after birth, likely due to enhanced neurological abnormalities affecting locomotive ability and proper feeding. These mice were also characterized by abnormal, under-developed cerebella. The majority of mice expressing only cyclin D2 died immediately after birth, with a small number able to survive up to 2 months. Again, the cause of their death was likely related to neurological defects leading to the aspiration of meconium into their lungs. Finally, development relying on cyclin D1 was terminated before birth, i.e., at 17.5–18.5 days of pregnancy. Analysis of surviving embryos revealed that they suffered from severe megaloblastic anemia [43]. Searching for the mechanisms allowing nearly normal development of the majority of tissues and organs in single-cyclin mice, we discovered upregulation of the remaining cyclin. This suggested the existence of a negative feedback loop in which a D-type cyclin that plays a key role in given tissue might repress the expression of the remaining ones. Interestingly, in tissues that failed to develop, e.g., cerebellum of D1$^{-/-}$D2$^{-/-}$ mice, the remaining cyclin (i.e., cyclin D3) was not upregulated. In the case of the cerebellum, this failure is caused by the inability of N-myc, which plays the crucial role in the proliferation of granule neuron precursors, to communicate with cell cycle machinery via cyclin D3. Thus, this result suggested the existence of transcription factor—cyclin D dependency [43]. The requirement of D1-associated kinase activity for cerebellar development was documented by the analysis of mice lacking cyclin D2 and expressing kinase-deficient cyclin D1. Such mice were characterized by severely retarded cerebellar development, leading to the conclusion that cyclin D-CDK4/CDK6 activity is necessary for morphogenesis of this organ [117]. Moreover, we also showed another feedback loop involving D-type cyclins, i.e., facilitation of cell cycle progression-mediated downregulation of p27kip1 levels [43].

Analysis of double knock-out mice suggested that either the presence of a single cyclin D allows nearly normal development of a majority of tissues and organs or proliferation of at least some cell types may occur in the absence of D-type cyclins. The latter scenario was proven by us by the generation of mice lacking all three D-type cyclins [109]. Cyclin D1$^{-/-}$D2$^{-/-}$D3$^{-/-}$ embryos developed until mid-gestation and died before 17.5 day of pregnancy. Detailed analyses of triple knock-out embryos revealed that indeed the majority of cell types can proliferate normally. Proliferative failure was limited to myocardial cells and hematopoietic stem cells making these lineages critically dependent on D-type cyclins [109]. Importantly, Cdk4 and Cdk6 double knock-out mouse embryos were also embryonic lethal, severely anemic, and displayed various defects in variety of hematopoietic lineages [144]. Interestingly, these embryos were able to survive a few days longer than cyclin D triple "knock-outs" most probably because D cyclins were able to interact and activate CDK2 [144]. Thus, cellular proliferation was shown to be possible without D-type cyclin-associated kinases, also suggested by the observation that in cyclin D2$^{-/-}$D3$^{-/-}$ embryos, in the presence of cyclin D1 only, CDK4 activity was not detectable [41]. Therefore, proliferation of only selected cell lines depends on cyclin D-CDK4/CDK6. At that time, however, it was uncertain if these cells fail to proliferate only because of their strict cell cycle requirements or because they require specific cyclin D functions, independent of CDK.

The fact that mouse embryos with acute ablation of all D-type cyclins failed to develop to term [109] made detailed studies of D cyclin function in peri- and post-natal development impossible. However, the development of the Cre-loxP system refining the conventional method of gene knock-out offered an excellent opportunity to avoid embryonic lethality and investigate consequences of acute ablation of a chosen gene in a tissue- and time-specific manner [129]. The technology of conditional gene knock-out is based on insertion of specific sequences (loxP or FRT) upstream and downstream of the target gene or gene fragment. Depending on the orientation of the sequences, the flanked region can be either irreversibly removed or inverted thanks to the activity or either Cre or FLP recombinase, respectively. By breeding mice carrying a floxed gene (f/f) with mice expressing Cre recombinase under the control of a tissue-specific promoter, tissue-/cell-specific deletion of a floxed gene is possible. Thus, using this method one can analyze the function of a selected gene in chosen tissue and at the chosen moment of embryonic or postnatal development, including adult organisms (reviewed in [193, 256]).

Based on the Cre-loxP system, conditional "knock-outs" of all three D-type cyclins were created that allowed a more precise test of the requirement for D-type cyclins in adult mice [38]. Using this technique conditional triple mouse mutants were generated by crossing "original" cyclin $D2^{-/-}$ mice with animals carrying conditionally modified cyclin D1- and D3-encoding genes. Next, these triple mutant mice were intercrossed with Mx1-Cre animals characterized by induced expression of Cre recombinase in hematopoietic cells [38]. A controlled shutdown of all D-type cyclins leads to abrupt disappearance of hematopoietic stem cells (HSCs), while the number of mature bone marrow cells remains unaffected, demonstrating that HSCs depend on D-type cyclins for their survival. The pro-survival function of D-type cyclins involves regulation of the death receptor Fas and its ligand FasL which, upon deletion of D-type cyclins, are strongly upregulated leading to the initiation of caspase-8-dependent apoptosis [38]. Thus, analysis of conditional knock-out of all D-type cyclins unraveled unexpected, non-cell cycle-related, functions of D-type cyclins in quiescent HSCs. This pro-survival role of D-type cyclins in adult hematopoietic cells was not demonstrated in the initial analysis of the "conventional" triple, i.e., cyclin D1, D2, and D3, knock-out model [109].

Conditional knock-out mice also allowed analyses of mice lacking a single D-type cyclin which, due to the embryonic or early postnatal lethality of mice, were impossible using traditionally derived knock-out mouse strains (e.g., [43, 109, 200]). Conditional deletion of cyclin D1 in liver proved that lack of this particular cyclin does not hamper liver development but uncovered the role of this protein in glucose metabolism regulation [127]. Lack of cyclin D1 does not induce changes in gluconeogenic gene expression and glycemia in fasting mice; however, in the re-fed state, it significantly increases expression of gluconeogenic genes, glycemia, glucose, and insulin intolerance. Thus, this study also revealed a cell cycle-unrelated function of cyclin D1, i.e., involvement in the regulation of nutrient and insulin signaling to regulate glucose metabolism. Importantly, inhibition of CDK4 activity fails to enhance this phenotype, suggesting that cyclin D1 alone mediates metabolic effects in the liver [127]. The conditional knock-out approach was also used in a study focusing on the cyclin-dependent kinases. Deletion of *Cdk4* and *Cdk2* results in lethality manifested shortly after birth [12]. Such mice die due to the failure of cardiac development. A decreased number of proliferating cardiomyocytes indicate that CDK4 and CDK2 play compensatory roles during heart development. Conditional ablation of *Cdk4* in *Cdk2*-null mice produces animals with no obvious abnormalities, proving that the function of adult tissues does not depend on CDK4 and CDK2 activity [12].

2.3.3.3   Quadruple and Quintuple Knock-Out Cells Enter the Stage

Generation of triple cyclin D knock-out mice and cell lines was followed by the derivation and analysis of the cells lacking either all G1 cyclins, i.e., cyclins D1, D2, D3, E1, and E2 [134], or those lacking cyclins E1, E2, A1, and A2 [102]. Surprisingly depletion of cyclins D and E did not block the proliferation of quintuple knock-out ES cells but completely prevent the proliferation of MEFs [134]. These ES cells, however, attenuated their pluripotent character and become prone to differentiate into trophectoderm. Further studies showed that G1 cyclin-dependent CDK activity is necessary to stabilize the pluripotency factors, such as Nanog, Sox2, and Oct4. Interestingly, ablation of G1/S cyclins, i.e., cyclins E and A, had no impact on MEFs [102].

2.3.3.4   What About Mice Deficient in CDK Inhibitors or pRb Family Members?

The goal of this chapter is to present the role of D cyclins and CDK4/CDK6 in cell cycle regulation during embryogenesis and cancer and also to describe some cell cycle-independent functions of these proteins. However, at least briefly, we would like to discuss the phenotypes of mice lacking some of the factors interacting with cyclin D-dependent kinases, i.e., CDK4/CDK6 inhibitors and pRb family members. The consequences of ablation of the expression of these proteins, resulting in the development of a variety of cancers, have been published in an enormous number of research articles, and it would be extremely difficult to present here a comprehensive summary. Thus, we will focus on studies describing the development and proper function of adult tissues.

The INK family of inhibitors includes $p16^{Ink4a}$, $p15^{Ink4b}$, $p18^{Ink4c}$, and $p19^{Ink4d}$ (e.g., [32, 79, 80, 83, 204]). Expression of $p16^{Ink4a}$ and $p15^{Ink4b}$ is detectable only in adult tissues and increases with age [257]. $p18^{Ink4c}$ and $p19^{Ink4d}$, on the other hand, are expressed in tissues of developing embryos as well as of adult animals [257, 259]. At the time of generation of $p16^{Ink4a}$ knock-out mice, it was not known that the *Ink4a* locus (CDKN2A) encodes not only $p16^{Ink4a}$ but also $p19^{Arf}$ [180]. Ablation of these two genes, however, did not result in obvious developmental abnormalities but promoted lymphomas and sarcomas [205]. Subsequently, knock-out mice were produced lacking $p16^{Ink4a}$ exclusively, and these animals developed almost normally. They were characterized by hyperplastic thymi, increased lymphocyte proliferation, and again high tumor incidence in keeping with a tumor suppressor function of $p16^{Ink4a}$ [110, 206, 207]. $p15^{Ink4b}$-null mice displayed hyperplastic lymph nodes and spleen, as well as extramedullary hematopoiesis, and also an increased proliferation rate of lymphocytes [119]. Lack of $p18^{Ink4c}$ alone also does not affect development. With age, however, $p18^{Ink4c}$-null mice become larger and reveal a hypoplastic pituitary gland and development of pituitary tumors, enlarged spleen, thymus, and other organs, as well as deregulated proliferation of epithelia, e.g., mammary gland epithelium [68, 119]. Deletion of genes encoding both $p15^{Ink4b}$ and $p18^{Ink4c}$ added some new phenotypes to those characteristic of the single knock-outs, i.e., double mutant mice suffer from enlarged testes and hyperplastic Langerhans islets [119]. Deficiency in $p19^{Ink4d}$ leads to male infertility due to testicular hyperplasia and hearing loss due to the malfunction of the auditory epithelium [35, 258, 260]. Therefore, the lack of inhibitors of cyclin D-CDK4/CDK6 complexes does not demonstrably impact embryonic development but in adult mice increases proliferation and leads to the development of hyperplasia of many organs, and eventually tumor development. On the other hand, ablation of cyclin D-CDK4/CDK6 substrates, i.e., pRb family members, results in much more severe phenotypes.

pRb, together with two other pRb-related proteins, namely, p107 and p130, is the first identified cyclin D-CDK4/CDK6 target [50]. Their phosphorylation and as a result inactivation are prerequisite for cell cycle progression since, as mentioned above, in the active state, they bind E2F transcription factors and prevent expression of crucial positive regulators of the cell cycle. During development, pRb is expressed starting from the peri-implantation stage of mouse embryo development (i.e., blastocyst) [93]. At later stages of development, all three pRb family proteins are specifically expressed in certain tissues [97]. The first studies focusing on Rb-null mice strongly suggested that this protein is indispensable for embryonic development. Knock-out mice died between 12 and 15 days of pregnancy due to severe anemia. They were also characterized by defects in lens development and massive cell death in the central (CNS) and peripheral nervous system (PNS) [44, 94, 124]. Generation of chimeric mice in which Rb-null cells were able to participate in the formation of many lineages, including the erythroid lineage, put in doubt a crucial role of this protein in hematopoiesis [138, 240]. Also, the neuronal apoptotic defects were not as obvious as described in the characterization of the phenotype of $Rb^{-/-}$ embryos. Generation of mice in which the Rb gene was conditionally deleted only in CNS, PNS, and lens revealed that CNS mutant tissues displayed ectopic S-phase entry but no apoptosis [65, 139].

Increased expression of hypoxia-inducible genes in Rb-null embryos suggested that observed apoptosis was induced by hypoxia [139]. This hypoxia, in turn, was thought to have resulted from placental malfunction. Experiments involving the "tetraploid complementation" technique allowing generation of mutant mice developing within wild-type placentas verified this notion [242]. Wu et al. proved that abnormal proliferation and differentiation of trophoblast cells prevented development of the labyrinth within the placenta which resulted in deficient nutrient and oxygen supply. These mice also died prematurely; however, they were able to develop to term, allowing observation of nearly normal development of the erythroid compartment and nervous system. Further studies showed that ablation of pRb in trophoblast stem cells resulted in abnormal trophoblast and placenta development [237].

The experiments described above, the creation of conditional mice and analyses of mice carrying *Rb* hypomorphic alleles, revealed the crucial role of pRb in embryonic myogenesis—muscle lacking pRb is characterized by hypoplastic myofibers [53, 242, 252]. Deletion of *Rb* in differentiating myoblasts resulted in apoptosis and failure to produce myotubes [88]. pRb's myogenic connection was also revealed during analysis of *p130* mutant mice. *p130*-null mice on a BALB/cJ background (characterized by reduced activity of p16^{Ink4a}) died in utero between days 11 and 13 of pregnancy due to defects in neuro- and myogenesis, i.e., reduced number of myocytes in the differentiating myotome [121]. The phenotype of *p107*-null mice is also influenced by the genetic background, i.e., BALB/cJ mutants were characterized by growth retardation and myeloid hyperplasia, but p107-deficient mice on a 129Sv/C57BL6 background displayed no obvious abnormalities [46, 122, 126]. *p107$^{-/-}$p130$^{-/-}$* animals were characterized by defects in chondrocyte proliferation and abnormal endochondral bone development [46]. Next, ablation of pRb either with p107 or p130 proved that these factors can substitute for each other, as the phenotype of either genetic combination is very similar—embryonic lethality occurs between 11 and 13 days of pregnancy due to liver and CNS abnormalities [126]. Finally, the consequence of deletion of genes encoding all three pRb family members analyzed in embryonic stem cells revealed that these proteins were crucial for successful differentiation and proper control of cellular proliferation [51, 196]. Again, as was the case with CDK inhibitors, deregulation of pRb protein expression led to tumor development, proving a crucial role of these cyclin D-CDK4/CDK6 regulators and substrates in the cell cycle control.

2.3.4 Cell Cycle-Independent Functions of Cyclins and CDKs

Many lines of evidence document that cyclins are involved in balancing proliferation and differentiation by impacting various tissue-specific transcription factors. Functions of cyclin D1 in this regard are the best studied so far, and the control of processes other than CDK4/CDK6 regulation is very well documented. Thus, it was shown that upregulation of cyclin D1 in cancer cells stimulates cellular migration by p27^{kip1} stabilization and also by impacting Rho protein function [130]. On the contrary, ablation of cyclin D1 negatively impacts cellular motility [170]. Cyclin D1 is also linked to DNA repair by data demonstrating that it can recruit RAD51 [131] and antagonize BRCA1-dependent repression of estrogen receptor α activity [234]. Further, cyclin D1 forms a complex with BRCA2, RAD51, and the Sp1 transcription factor [174, 228], interacts with PCNA [152, 243] and replication factor C (RFC) [233], all of which are also involved in DNA repair. In addition to the aforementioned functions, cyclin D1 involvement in the regulation of transcription is unquestionable. For example, cyclin D1 was shown to compete with androgen receptor for p300/CBP-associated factor (P/CAF) binding [189] and to inhibit the function of peroxisome proliferator-activated receptor gamma (PPARgamma) [235], to interact with transcription factors such as myb-like binding protein (DMP1) [92], repress STAT3 [20, 21] and inhibit NeuroD function [135, 185]. Moreover, both D1 and D2 cyclins inhibit transcription activated via the v-Myb DNA-binding domain [69].

The identification of additional novel cyclin D1 roles was possible due to the generation of knock-in mice expressing proteins labeled with such tags as Flag or hemagglutinin (HA). This approach was initially used by Bienvenu et al. who generated transgenic mice expressing Flag- and HA-tagged cyclin D1 [22]. By sequential immunoaffinity purification using anti-Flag and anti-HA antibodies, followed by repeated rounds of high-throughput mass spectrometry, novel cyclin D1-interacting proteins were identified. Among cyclin D1 interactors identified were known cell cycle partners, such as CDK4 and CDK6, and those less typical, such as CDK1, CDK2, CDK5, and CDK11. This study also confirmed involvement of cyclin D1 in the regulation of transcription—it was shown to bind to promoter regions of more than 900 genes [22]. Importantly, this approach revealed the mechanism leading to the retinal phenotype characteristic of D1-deficient mice. In retinas, cyclin D1 physically binds and recruits CBP histone acetyltransferase to the Notch1 upstream regulatory region [22]. In the absence of cyclin D1, acetylation of histones was decreased, resulting in transcriptional repression of the targeted gene, i.e., Notch1. Cyclin D1 transcriptional function in the development of other tissues and cancer formation will be the next major goal of many research efforts using this approach. In the meantime, protein interactome analyses of human cancers proved cyclin D1 interaction with DNA repair proteins, including RAD51 [99, 100]. Thus, the generation of knock-in mice carrying genes encoding tagged proteins provided a unique chance to uncover a whole new world of previously unappreciated protein functions.

2.4   Cyclin D- and CKD4/6-Deficient Mice Versus Cancer

Oncogenic roles of D-type cyclins and their CDK partners are widely documented by analyses showing that these proteins are overexpressed in a variety of tumors and by experiments involving either their overexpression or elimination [142, 143, 168]. Aberrant cyclin D1 expression is observed in a wide spectrum of human cancers, such as colorectal cancer, uterine cancer, malignant melanoma, squamous cell carcinoma of head and neck, astrocytoma, non-small-cell lung cancer, soft tissue sarcoma, and others [14, 17, 67, 123, 140, 151, 198]. Importantly, breast cancer is perhaps the best documented malignancy involving cyclin D1. Approximately 15–20% of mammary tumors contain amplification of the *CCND1* gene whereas its overexpression is detected in over 50% [14, 72, 116, 262]. Interestingly, overexpression of cyclin D1 is more common than can be explained by gene altera-tion. Therefore, other mechanisms such as deregulation of mitogenic signaling path-ways or aberrant proteolytic degradation must underlie cyclin D1 overexpression. Indeed, elevated levels of cyclin D1 protein were observed in the absence of increased mRNA reflecting a defect in its proteolysis [194]. This effect was confirmed in trans-genic mice expressing phosphorylation-deficient cyclin D1 under the control of the tissue (i.e., mammary gland)-specific MMTV promoter. Disruption of cyclin D1 phosphorylation led to the accumulation of the protein in the nucleus, prevented its cytoplasmic proteolysis and accelerated mammary carcinogenesis [132].

MMTV-driven expression in transgenic mice has facilitated analysis of mam-mary gland-specific expression of various oncogenes, including cyclins, associated kinases, inhibitors, Ras, Myc, and others [227]. In 1994, MMTV-cyclin D1 mice were shown to develop mammary adenocarcinomas within 22 months of age [236]. The relatively late occurrence of these mammary tumors suggests involvement of other oncogenic pathways. Interestingly, intercrossing MMTV-cyclin D1 with $p53^{+/-}$ mice did not result in mammary neoplasia [84]. In mice heterozygous for p53 deficiency and simultaneously carrying the MMTV-cyclin D1 transgene, only tumors typical for p53-deficient mice developed, and interestingly, their growth was significantly accelerated by cyclin D1 overexpression. Surprisingly, mammary tumors were not observed. More rapid development of non-mammary tumors in MMTV-cyclin D1/$p53^{+/-}$, as compared with $p53^{+/-}$, raise the possibility that p53 inactivation might complement or cooperate with cyclin D1 deregulation during the development of some types of non-mammary tumors.

The connection between cyclin D1, and also D2 and D3, and tumorigenesis was strengthened by the analyses of mice, or cells derived from them, that lacked single, two, or all three D-type cyclins or their CDK partners. Lack of cyclin D1 prevents not only physiological but also pathological proliferation of mammary gland epithe-lium. Yu et al. revealed that breast tumors arising in MMTV-ras and MMTV-neu mice expressed almost exclusively cyclin D1, very low levels of cyclin D3, and no cyclin D2 [249]. In contrast, several tumors arising in MMTV-Wnt-1 and MMTV-myc females expressed, in addition to cyclin D1, also high levels of D2. Importantly, all tumors arose from luminal epithelial cells, indicating that, in mammary epithelial cells, Ras and Neu oncogenes communicate with the cell cycle machinery through cyclin D1, whereas Wnt-1 and Myc can signal through other targets. Therefore, therapies involving cyclin D1 inhibition might be highly selective in shutting off the growth of human breast cancers, particularly those characterized by amplification and/or overexpression of c-Neu (ErbB-2, HER-2). This hypothesis was recently challenged in a genetic mouse model that allows controlled expression of cyclin D1 in progressing mammary tumors [255]. Zhang observed that cyclin D1 defi-ciency delayed the development of tumors; however, it did not protect against ErbB2-driven mammary carcinogenesis as previously reported [249]. Moreover, in the absence of cyclin D1, cyclin D3 was upregulated. Knockdown of cyclin D3 in tumor-derived cells lacking cyclin D1$^{-/-}$ resulted in significant tumor growth impair-ment in comparison to cells expressing cyclin D3. It is, therefore, possible that only the combined inhibition of cyclin D1 and D3 might serve as an effective strategy for breast cancer therapy. Further studies demonstrated that cyclin D1 absence suppressed Neu- and mutant Neu (activated c-neu)-driven mammary tumor formation confirming that cyclin D1 is required for the Neu-driven signal transduction pathway [25]. Interestingly, no significant changes in either cyclin D2 or cyclin D3 expression were detected in MMTV-c-neu/cyclin D1$^{-/-}$-derived mammary tumors. However, increased levels of cyclin E and higher activity of cyclin E-CDK2 complexes were demonstrated. Thus, Bowe et al. suggested that neither cyclin D2 nor D3 compensate for the absence of cyclin D1 to promote the oncogenic potential of Neu [25]. The above discrepancies were addressed by Choi et al. who created conditional cyclin D1 and D3 knock-out mice allowing acute ablation of individual cyclins [37]. Contrary to what was presented by Zhang et al., induced ablation of cyclin D1 in the whole body, including ErbB2-driven mammary carcinomas, resulted in cessation of tumor progression [37].

Mice expressing a mutated form of cyclin D1 proved that cyclin-associated CDK activity is crucial for oncogene-induced breast cancer development [117, 247]. Knock-in mice expressing kinase-deficient cyclin D1-CDK4 complexes are resistant to mammary carcinomas triggered by ErbB-2 [117]. Also, analyses of CDK4-deficient mice confirmed the role of CDK4 in breast cancer [187, 188, 251]. Therefore, it was not surprising that administration of PD0332991, a specific and potent inhibitor of cyclin D-CDK4/CDK6 kinases, halted the progression of breast cancers [37]. Interestingly, cyclin E was shown to be able to replace the function of cyclin D1 in Wnt-induced tumors [70], however, CDK4 function seems to be unreplaceable by CDK6 [188]. Remarkably, cyclin D1-CDK2 complexes were present in mammary carcinoma cells; hence, they might be an additional factor contributing to the oncogenic effects of cyclin D1 overexpression [223]. Indeed, transgenic mice expressing a cyclin D1-CDK2 fusion protein under the control of the MMTV promoter developed breast tumors [49].

Cyclin D1 gene amplification was demonstrated in breast cancers in which *CCND1* overexpression was linked to estrogen and progesterone receptor status (reviewed in [178]). This connection is attributed to cyclin D1 regulation by estrogen (ER) and interaction of cyclin D1 with ER coactivators to activate estrogen receptor binging elements (ERE) in a CDK4−/CDK6-independent manner [169, 197, 263]. Furthermore, cyclin D1 was shown to regulate progesterone receptor (PR) expression, through an estrogen- and cyclin D1-responsive enhancer localized on the 3'UTR [246]. Loss of cyclin D1 led to decreased PR mRNA levels in mammary glands. In addition, a higher risk of development of tumors that express estrogen receptor is associated with elevated prolactin (PRL) and PRL receptor (PRLR) levels—both critical for epithelial proliferation during development and pregnancy [222, 232]. Cyclin D1$^{-/-}$ mouse epithelial cells fail to proliferate in response to prolactin [26]. Although deletion of cyclin D1 in transgenic mice overexpressing PRL markedly decreased tumor incidence, cyclin D1$^{-/-}$ females overexpressing PRL developed significantly more preneoplastic lesions than D1$^{-/-}$ females [6]. Interestingly, tumors that formed in this background exhibited elevated levels of cyclin D3 and a squamous histotype similar to those that developed in MMTV-cyclin D3 mice [179].

Cyclin D1-deficient mice were also shown to be "resistant" to other cancers. For example, they do not develop Ras-triggered skin papillomas [190] or intestinal polyps in the Apc^{min} background [89]. Extending results with cyclin D1 overexpressing or null mice, the involvement of other D cyclins in carcinogenesis was also documented. Despite the fact that cyclin D2 was shown to be a direct target of Myc [24, 177], much less attention has been devoted to investigation of cyclin D2 involvement in breast cancer. Clinical data demonstrate, however, that cyclin D2 is absent in breast cancer cell lines and tumors [28, 60]. Mice lacking cyclin D2 are characterized by reduced susceptibility to gonadal tumors [29] and insensitivity to BCR/ABL-driven transformation [95] and, similarly to D1-deficient animals, to Apc^{min}-induced formation of intestinal polyps [47]. Aberrant accumulation of cyclin D3 was also documented in a subset of breast carcinomas [13, 194]. As mentioned above, ablation of cyclin D3 in cyclin D1$^{-/-}$ mice further reduces mammary tumor development. Lack of cyclin D3 was shown to result in delayed development of thymomas caused by p56^{lck} and resistance to Notch-driven leukemias (acute lymphoblastic leukemia, T-ALL) [211], and acute ablation of cyclin D3 in abnormal CD4$^+$CD8$^+$ cells blocked the development of Notch1-driven T-ALL in vivo [37].

Although all D-type cyclins are highly related and are expressed in a largely overlapping fashion, it is clear that there are differences in their specificity to transmit specific oncogenic signals to the cell cycle machinery. Thus, requirement for D-type cyclins in oncogenic transformation was also tested using mouse embryonic fibroblasts lacking two or all three D-type cyclins [109, 250]. Each of the D-type cyclins is certainly sufficient to mediate the action of such oncogenes as Ras and c-Myc [250]. However, triple knock-out fibroblasts are resistant to the action of Ras, c-Myc, or Ras combined with c-Myc, dnp53, or E1A [109]. Also CDK4-deficient cells are unaffected by Ras and dnp53 [181, 261]. Further, CDK4 deficiency in mice resulted in decreased incidence of skin tumors [191] as well as Myc-induced tumors in the oral mucosa [160]. Similarly, CDK6-deficient mice were shown to be resistant to Akt-driven lymphoma [86] as well as BCR-ABL^{p210}-driven leukemia [203].

Current development of high-throughput platforms allows to study interactomes of various factors involved in oncogenesis, such as cyclin D1 or CDK4 [100, 101, 172, 176]. Results of such analyses document cell cycle-dependent functions of cyclins and CDKs as well as reveal their non-canonical properties (for the summary, see [90]). Such studies are of the vital importance for the development of future therapeutic approaches.

2.5 Concluding Remarks

Deletion of the genes encoding D-type cyclins and their partners provided valuable clues about their role in embryonic development and in cell cycle progression of different cell types. Consequently, observing the characteristics of mice lacking these genes has provided copious information that can be used to better understand D cyclin contributions to cancer formation. Taking advantage of conditional knock-out mice lacking one or more D cyclins, it has been possible to determine if a particular D cyclin is required at different developmental stages or for tumor initiation and maintenance. Since aberrant expression of cell cycle regulators is very frequent in tumorigenesis, it is of outmost importance to test if these proteins could be potentially targeted in various therapeutic approaches. Another burning problem that can be addressed using conditional mouse models is whether therapeutic targeting of those proteins will have negative consequences in tumor-free organs. Furthermore, dissection of the mammalian cell cycle machinery, including uncovering novel cyclin D1 roles, is possible based on the generation of variety of knock-in mice, including animals expressing tagged proteins. The exploration of such new tools has already brought surprising results although the process has just begun. Thus, the discovery of novel D cyclin roles in mechanisms regulating normal and tumor cell cycles is ongoing, and these studies will be invaluable in extending the understanding and application of current therapies targeting D-cyclin-dependent kinases.

Acknowledgments We want to thank Piotr Sicinski for being our mentor in the cyclin universe. We are also grateful to Katarzyna Koziak and Phil Hinds for their help at the very final stage of the manuscript preparation. During the preparation of this chapter, IK was supported by the Medical University of Warsaw statutory grant 1M15/N/2015 and the National Centre for Research and Development (NCBR) grant STRATEGMED3/307326/6/NCBR/2017 and MAC by the Faculty of Biology University of Warsaw funding 501/56/169600 and the National Science Centre Poland (NCN) grant 2012/05/N/NZ3/00314.

References

1. Abella A, Dubus P, Malumbres M, Rane SG, Kiyokawa H, Sicard A, Vignon F, Langin D, Barbacid M, Fajas L. Cdk4 promotes adipogenesis through PPARgamma activation. Cell Metab. 2005;2:239–49.
2. Aguzzi A, Kiess M, Rued D, Hamel PA. Cyclins D1, D2 and D3 are expressed in distinct tissues during mouse embryogenesis. Transgenics. 1996;2:29–39.
3. Albanese C, Johnson J, Watanabe G, Eklund N, Vu D, Arnold A, Pestell RG. Transforming p21ras mutants and c-Ets-2 activate the cyclin D1 promoter through distinguishable regions. J Biol Chem. 1995;270:23589–97.
4. Anders L, Ke N, Hydbring P, Choi YJ, Widlund HR, Chick JM, Zhai HL, Vidal M, Gygi SP, Braun P, et al. A systematic screen for CDK4/6 substrates links FOXM1 phosphorylation to senescence suppression in cancer cells. Cancer Cell. 2011;20:620–34.
5. Ansorg A, Witte OW, Urbach A. Age-dependent kinetics of dentate gyrus neurogenesis in the absence of cyclin D2. BMC Neurosci. 2012;13:46.
6. Asher JM, O'Leary KA, Rugowski DE, Arendt LM, Schuler LA. Prolactin promotes mammary pathogenesis independently from cyclin D1. Am J Pathol. 2012;181:294–302.
7. Atanasoski S, Shumas S, Dickson C, Scherer SS, Suter U. Differential cyclin D1 requirements of proliferating Schwann cells during development and after injury. Mol Cell Neurosci. 2001;18:581–92.
8. Atanasoski S, Boentert M, De Ventura L, Pohl H, Baranek C, Beier K, Young P, Barbacid M, Suter U. Postnatal Schwann cell proliferation but not myelination is strictly and uniquely dependent on cyclin-dependent kinase 4 (cdk4). Mol Cell Neurosci. 2008;37:519–27.

9. Bagui TK, Jackson RJ, Agrawal D, Pledger WJ. Analysis of cyclin D3-cdk4 complexes in fibroblasts expressing and lacking p27(kip1) and p21(cip1). Mol Cell Biol. 2000;20:8748–57.
10. Baker GL, Landis MW, Hinds PW. Multiple functions of D-type cyclins can antagonize pRb-mediated suppression of proliferation. Cell Cycle. 2005;4:330–8.
11. Balakier H, Czolowska R. Cytoplasmic control of nuclear maturation in mouse oocytes. Exp Cell Res. 1977;110:466–9.
12. Barriere C, Santamaria D, Cerqueira A, Galan J, Martin A, Ortega S, Malumbres M, Dubus P, Barbacid M. Mice thrive without Cdk4 and Cdk2. Mol Oncol. 2007;1:72–83.
13. Bartkova J, Zemanova M, Bartek J. Abundance and subcellular localisation of cyclin D3 in human tumours. Int J Cancer. 1996;65:323–7.
14. Bartkova J, Lukas J, Strauss M, Bartek J. Cyclin D1 oncoprotein aberrantly accumulates in malignancies of diverse histogenesis. Oncogene. 1995;10:775–8.
15. Bartkova J, Lukas J, Strauss M, Bartek J. Cyclin D3: requirement for G1/S transition and high abundance in quiescent tissues suggest a dual role in proliferation and differentiation. Oncogene. 1998;17:1027–37.
16. Bartkova J, Rajpert-De Meyts E, Skakkebaek NE, Bartek J. D-type cyclins in adult human testis and testicular cancer: relation to cell type, proliferation, differentiation, and malignancy. J Pathol. 1999;187:573–81.
17. Bartkova J, Lukas J, Muller H, Strauss M, Gusterson B, Bartek J. Abnormal patterns of D-type cyclin expression and G1 regulation in human head and neck cancer. Cancer Res. 1995;55:949–56.
18. Bates S, Bonetta L, MacAllan D, Parry D, Holder A, Dickson C, Peters G. CDK6 (PLSTIRE) and CDK4 (PSK-J3) are a distinct subset of the cyclin-dependent kinases that associate with cyclin D1. Oncogene. 1994;9:71–9.
19. Beumer TL, Roepers-Gajadien HL, Gademan IS, Kal HB, de Rooij DG. Involvement of the D-type cyclins in germ cell proliferation and differentiation in the mouse. Biol Reprod. 2000;63:1893–8.
20. Bienvenu F, Gascan H, Coqueret O. Cyclin D1 represses STAT3 activation through a Cdk4-independent mechanism. J Biol Chem. 2001;276:16840–7.
21. Bienvenu F, Barre B, Giraud S, Avril S, Coqueret O. Transcriptional regulation by a DNA-associated form of cyclin D1. Mol Biol Cell. 2005;16:1850–8.
22. Bienvenu F, Jirawatnotai S, Elias JE, Meyer CA, Mizeracka K, Marson A, Frampton GM, Cole MF, Odom DT, Odajima J, et al. Transcriptional role of cyclin D1 in development revealed by a genetic-proteomic screen. Nature. 2010;463:374–8.
23. Blain SW, Montalvo E, Massague J. Differential interaction of the cyclin-dependent kinase (Cdk) inhibitor p27Kip1 with cyclin A-Cdk2 and cyclin D2-Cdk4. J Biol Chem. 1997;272:25863–72.
24. Bouchard C, Thieke K, Maier A, Saffrich R, Hanley-Hyde J, Ansorge W, Reed S, Sicinski P, Bartek J, Eilers M. Direct induction of cyclin D2 by Myc contributes to cell cycle progression and sequestration of p27. EMBO J. 1999;18:5321–33.
25. Bowe DB, Kenney NJ, Adereth Y, Maroulakou IG. Suppression of Neu-induced mammary tumor growth in cyclin D1 deficient mice is compensated for by cyclin E. Oncogene. 2002;21:291–8.
26. Brisken C, Ayyannan A, Nguyen C, Heineman A, Reinhardt F, Tan J, Dey SK, Dotto GP, Weinberg RA. IGF-2 is a mediator of prolactin-induced morphogenesis in the breast. Dev Cell. 2002;3:877–87.
27. Brown NE, Jeselsohn R, Bihani T, Hu MG, Foltopoulou P, Kuperwasser C, Hinds PW. Cyclin D1 activity regulates autophagy and senescence in the mammary epithelium. Cancer Res. 2012;72:6477–89.
28. Buckley MF, Sweeney KJ, Hamilton JA, Sini RL, Manning DL, Nicholson RI, de Fazio A, Watts CK, Musgrove EA, Sutherland RL. Expression and amplification of cyclin genes in human breast cancer. Oncogene. 1993;8:2127–33.

29. Burns KH, Agno JE, Sicinski P, Matzuk MM. Cyclin D2 and p27 are tissue-specific regulators of tumorigenesis in inhibin alpha knockout mice. Mol Endocrinol. 2003;17:2053–69.
30. Carthon BC, Neumann CA, Das M, Pawlyk B, Li T, Geng Y, Sicinski P. Genetic replacement of cyclin D1 function in mouse development by cyclin D2. Mol Cell Biol. 2005;25:1081–8.
31. Cenciarelli C, De Santa F, Puri PL, Mattei E, Ricci L, Bucci F, Felsani A, Caruso M. Critical role played by cyclin D3 in the MyoD-mediated arrest of cell cycle during myoblast differentiation. Mol Cell Biol. 1999;19:5203–17.
32. Chan FK, Zhang J, Cheng L, Shapiro DN, Winoto A. Identification of human and mouse p19, a novel CDK4 and CDK6 inhibitor with homology to p16ink4. Mol Cell Biol. 1995;15:2682–8.
33. Chen B, Pollard JW. Cyclin D2 compensates for the loss of cyclin D1 in estrogen-induced mouse uterine epithelial cell proliferation. Mol Endocrinol. 2003;17:1368–81.
34. Chen J, Larochelle S, Li X, Suter B. Xpd/Ercc2 regulates CAK activity and mitotic progression. Nature. 2003;424:228–32.
35. Chen P, Zindy F, Abdala C, Liu F, Li X, Roussel MF, Segil N. Progressive hearing loss in mice lacking the cyclin-dependent kinase inhibitor Ink4d. Nat Cell Biol. 2003;5:422–6.
36. Chen Z, Duan RS, Zhu Y, Folkesson R, Albanese C, Winblad B, Zhu J. Increased cyclin E expression may obviate the role of cyclin D1 during brain development in cyclin D1 knockout mice. J Neurochem. 2005;92:1281–4.
37. Choi YJ, Li X, Hydbring P, Sanda T, Stefano J, Christie AL, Signoretti S, Look AT, Kung AL, von Boehmer H, et al. The requirement for cyclin D function in tumor maintenance. Cancer Cell. 2012;22:438–51.
38. Choi YJ, Saez B, Anders L, Hydbring P, Stefano J, Bacon NA, Cook C, Kalaszczynska I, Signoretti S, Young RA, et al. D-cyclins repress apoptosis in hematopoietic cells by controlling death receptor Fas and its ligand FasL. Dev Cell. 2014;30:255–67.
39. Chow YH, Zhu XD, Liu L, Schwartz BR, Huang XZ, Harlan JM, Schnapp LM. Role of Cdk4 in lymphocyte function and allergen response. Cell Cycle. 2010;9:4922–30.
40. Ciemerych MA, Sicinski P. Cell cycle in mouse development. Oncogene. 2005;24:2877–98.
41. Ciemerych MA, Yu Q, Szczepanska K, Sicinski P. CDK4 activity in mouse embryos expressing a single D-type cyclin. Int J Dev Biol. 2008;52:299–305.
42. Ciemerych MA, Archacka K, Grabowska I, Przewozniak M. Cell cycle regulation during proliferation and differentiation of mammalian muscle precursor cells. Results Probl Cell Differ. 2011;53:473–527.
43. Ciemerych MA, Kenney AM, Sicinska E, Kalaszczynska I, Bronson RT, Rowitch DH, Gardner H, Sicinski P. Development of mice expressing a single D-type cyclin. Genes Dev. 2002;16:3277–89.
44. Clarke AR, Maandag ER, van Roon M, van der Lugt NM, van der Valk M, Hooper ML, Berns A, te Riele H. Requirement for a functional Rb-1 gene in murine development. Nature. 1992;359:328–30.
45. Cobrinik D. Pocket proteins and cell cycle control. Oncogene. 2005;24:2796–809.
46. Cobrinik D, Lee MH, Hannon G, Mulligan G, Bronson RT, Dyson N, Harlow E, Beach D, Weinberg RA, Jacks T. Shared role of the pRB-related p130 and p107 proteins in limb development. Genes Dev. 1996;10:1633–44.
47. Cole AM, Myant K, Reed KR, Ridgway RA, Athineos D, Van den Brink GR, Muncan V, Clevers H, Clarke AR, Sicinski P, et al. Cyclin D2-cyclin-dependent kinase 4/6 is required for efficient proliferation and tumorigenesis following Apc loss. Cancer Res. 2010;70:8149–58.
48. Cooper AB, Sawai CM, Sicinska E, Powers SE, Sicinski P, Clark MR, Aifantis I. A unique function for cyclin D3 in early B cell development. Nat Immunol. 2006;7:489–97.
49. Corsino P, Davis B, Law M, Chytil A, Forrester E, Norgaard A, Teoh N, Law B. Tumors initiated by Constitutive Cdk2 activation exhibit transforming growth factor A resistance and acquire paracrine mitogenic stimulation during progression. Cancer Research. 2007;67(7):3135–44.

50. Dannenberg JH, te Riele HP. The retinoblastoma gene family in cell cycle regulation and suppression of tumorigenesis. Results Probl Cell Differ. 2006;42:183–225.
51. Dannenberg JH, van Rossum A, Schuijff L, te Riele H. Ablation of the retinoblastoma gene family deregulates G(1) control causing immortalization and increased cell turnover under growth-restricting conditions. Genes Dev. 2000;14:3051–64.
52. Das G, Clark AM, Levine EM. Cyclin D1 inactivation extends proliferation and alters histogenesis in the postnatal mouse retina. Dev Dyn. 2012;241:941–52.
53. de Bruin A, Wu L, Saavedra HI, Wilson P, Yang Y, Rosol TJ, Weinstein M, Robinson ML, Leone G. Rb function in extraembryonic lineages suppresses apoptosis in the CNS of Rb-deficient mice. Proc Natl Acad Sci U S A. 2003;100:6546–51.
54. Dimova DK, Dyson NJ. The E2F transcriptional network: old acquaintances with new faces. Oncogene. 2005;24:2810–26.
55. Dowdy SF, Hinds PW, Louie K, Reed SI, Arnold A, Weinberg RA. Physical interaction of the retinoblastoma protein with human D cyclins. Cell. 1993;73:499–511.
56. Ericson KK, Krull D, Slomiany P, Grossel MJ. Expression of cyclin-dependent kinase 6, but not cyclin-dependent kinase 4, alters morphology of cultured mouse astrocytes. Mol Cancer Res. 2003;1:654–64.
57. Ericson KK, Gocheva V, Slomiany P, Udoeyop I, Grossel MJ. Differences in Cdk4 and Cdk6 function in primary mouse astrocytes. Mol Biol Cell. 2002;13:438a.
58. Evans MJ, Kaufman MH. Establishment in culture of pluripotential cells from mouse embryos. Nature. 1981;292:154–6.
59. Evans T, Rosenthal ET, Youngblom J, Distel D, Hunt T. Cyclin: a protein specified by maternal mRNA in sea urchin eggs that is destroyed at each cleavage division. Cell. 1983;33:389–96.
60. Evron E, Umbricht CB, Korz D, Raman V, Loeb DM, Niranjan B, Buluwela L, Weitzman SA, Marks J, Sukumar S. Loss of cyclin D2 expression in the majority of breast cancers is associated with promoter hypermethylation. Cancer Res. 2001;61:2782–7.
61. Faast R, White J, Cartwright P, Crocker L, Sarcevic B, Dalton S. Cdk6-cyclin D3 activity in murine ES cells is resistant to inhibition by p16(INK4a). Oncogene. 2004;23:491–502.
62. Fantl V, Stamp G, Andrews A, Rosewell I, Dickson C. Mice lacking cyclin D1 are small and show defects in eye and mammary gland development. Genes Dev. 1995;9:2364–72.
63. Fantl V, Edwards PA, Steel JH, Vonderhaar BK, Dickson C. Impaired mammary gland development in Cyl-1(−/−) mice during pregnancy and lactation is epithelial cell autonomous. Dev Biol. 1999;212:1–11.
64. Fantl V, Creer A, Dillon C, Bresnick J, Jackson D, Edwards P, Rosewell I, Dickson C. Fibroblast growth factor signalling and cyclin D1 function are necessary for normal mammary gland development during pregnancy. A transgenic mouse approach. Adv Exp Med Biol. 2000;480:1–7.
65. Ferguson KL, Vanderluit JL, Hebert JM, McIntosh WC, Tibbo E, MacLaurin JG, Park DS, Wallace VA, Vooijs M, McConnell SK, et al. Telencephalon-specific Rb knockouts reveal enhanced neurogenesis, survival and abnormal cortical development. EMBO J. 2002;21:3337–46.
66. Fisher RP. Secrets of a double agent: CDK7 in cell-cycle control and transcription. J Cell Sci. 2005;118:5171–80.
67. Florenes VA, Faye RS, Maelandsmo GM, Nesland JM, Holm R. Levels of cyclin D1 and D3 in malignant melanoma: deregulated cyclin D3 expression is associated with poor clinical outcome in superficial melanoma. Clin Cancer Res. 2000;6:3614–20.
68. Franklin DS, Godfrey VL, Lee H, Kovalev GI, Schoonhoven R, Chen-Kiang S, Su L, Xiong Y. CDK inhibitors p18(INK4c) and p27(Kip1) mediate two separate pathways to collaboratively suppress pituitary tumorigenesis. Genes Dev. 1998;12:2899–911.
69. Ganter B, Fu SL, Lipsick JS. D-type cyclins repress transcriptional activation by the v-Myb but not the c-Myb DNA-binding domain. EMBO J. 1998;17:255–68.
70. Geng Y, Whoriskey W, Park MY, Bronson RT, Medema RH, Li T, Weinberg RA, Sicinski P. Rescue of cyclin D1 deficiency by knockin cyclin E. Cell. 1999;97:767–77.

71. Georgia S, Bhushan A. Beta cell replication is the primary mechanism for maintaining post-natal beta cell mass. J Clin Invest. 2004;114:963–8.
72. Gillett C, Fantl V, Smith R, Fisher C, Bartek J, Dickson C, Barnes D, Peters G. Amplification and overexpression of cyclin D1 in breast cancer detected by immunohistochemical staining. Cancer Res. 1994;54:1812–7.
73. Glickstein SB, Alexander S, Ross ME. Differences in cyclin D2 and D1 protein expression distinguish forebrain progenitor subsets. Cereb Cortex. 2007;17:632–42.
74. Glickstein SB, Monaghan JA, Koeller HB, Jones TK, Ross ME. Cyclin D2 is critical for intermediate progenitor cell proliferation in the embryonic cortex. J Neurosci. 2009;29:9614–24.
75. Glickstein SB, Moore H, Slowinska B, Racchumi J, Suh M, Chuhma N, Ross ME. Selective cortical interneuron and GABA deficits in cyclin D2-null mice. Development. 2007;134:4083–93.
76. Gopinathan L, Ratnacaram CK, Kaldis P. Established and novel Cdk/cyclin complexes regulating the cell cycle and development. Results Probl Cell Differ. 2011;53:365–89.
77. Grossel MJ, Hinds PW. From cell cycle to differentiation – an expanding role for Cdk6. Cell Cycle. 2006;5:266–70.
78. Grossel MJ, Hinds PW. Beyond the cell cycle: a new role for cdk6 in differentiation. J Cell Biochem. 2006;97:485–93.
79. Guan KL, Jenkins CW, Li Y, O'Keefe CL, Noh S, Wu X, Zariwala M, Matera AG, Xiong Y. Isolation and characterization of p19INK4d, a p16-related inhibitor specific to CDK6 and CDK4. Mol Biol Cell. 1996;7:57–70.
80. Hannon GJ, Beach D. p15INK4B is a potential effector of TGF-beta-induced cell cycle arrest. Nature. 1994;371:257–61.
81. Hartwell LH, Culotti J, Pringle JR, Reid BJ. Genetic control of the cell division cycle in yeast. Science. 1974;183:46–51.
82. Hindley J, Phear GA. Sequence of the cell division gene CDC2 from Schizosaccharomyces pombe; patterns of splicing and homology to protein kinases. Gene. 1984;31:129–34.
83. Hirai H, Roussel MF, Kato JY, Ashmun RA, Sherr CJ. Novel INK4 proteins, p19 and p18, are specific inhibitors of the cyclin D-dependent kinases CDK4 and CDK6. Mol Cell Biol. 1995;15:2672–81.
84. Hosokawa Y, Papanikolaou A, Cardiff RD, Yoshimoto K, Bernstein M, Wang TC, Schmidt EV, Arnold A. In vivo analysis of mammary and non-mammary tumorigenesis in MMTV-cyclin D1 transgenic mice deficient in p53. Transgenic Res. 2001;10:471–8.
85. Hu MG, Deshpande A, Schlichting N, Hinds EA, Mao C, Dose M, Hu GF, Van Etten RA, Gounari F, Hinds PW. CDK6 kinase activity is required for thymocyte development. Blood. 2011;117:6120–31.
86. Hu MG, Deshpande A, Enos M, Mao DQ, Hinds EA, Hu GF, Chang R, Guo ZY, Dose M, Mao CC, et al. A requirement for cyclin-dependent kinase 6 in thymocyte development and tumorigenesis. Cancer Res. 2009;69:810–8.
87. Huard JM, Forster CC, Carter ML, Sicinski P, Ross ME. Cerebellar histogenesis is disturbed in mice lacking cyclin D2. Development. 1999;126:1927–35.
88. Huh MS, Parker MH, Scime A, Parks R, Rudnicki MA. Rb is required for progression through myogenic differentiation but not maintenance of terminal differentiation. J Cell Biol. 2004;166:865–76.
89. Hulit J, Wang C, Li Z, Albanese C, Rao M, Di Vizio D, Shah S, Byers SW, Mahmood R, Augenlicht LH, et al. Cyclin D1 genetic heterozygosity regulates colonic epithelial cell differentiation and tumor number in ApcMin mice. Mol Cell Biol. 2004;24:7598–611.
90. Hydbring P, Malumbres M, Sicinski P. Non-canonical functions of cell cycle cyclins and cyclin-dependent kinases. Nat Rev Mol Cell Biol. 2016;17:280–92.
91. Inaba T, Matsushime H, Valentine M, Roussel MF, Sherr CJ, Look AT. Genomic organization, chromosomal localization, and independent expression of human cyclin D genes. Genomics. 1992;13:565–74.
92. Inoue K, Sherr CJ. Gene expression and cell cycle arrest mediated by transcription factor DMP1 is antagonized by D-type cyclins through a cyclin-dependent-kinase-independent mechanism. Mol Cell Biol. 1998;18:1590–600.

93. Iwamori N, Naito K, Sugiura K, Tojo H. Preimplantation-embryo-specific cell cycle regulation is attributed to the low expression level of retinoblastoma protein. FEBS Lett. 2002;526:119–23.
94. Jacks T, Fazeli A, Schmitt EM, Bronson RT, Goodell MA, Weinberg RA. Effects of an Rb mutation in the mouse. Nature. 1992;359:295–300.
95. Jena N, Deng M, Sicinska E, Sicinski P, Daley GQ. Critical role for cyclin D2 in BCR/ABL-induced proliferation of hematopoietic cells. Cancer Res. 2002;62:535–41.
96. Jeselsohn R, Brown NE, Arendt L, Klebba I, Hu MG, Kuperwasser C, Hinds PW. Cyclin D1 kinase activity is required for the self-renewal of mammary stem and progenitor cells that are targets of MMTV-ErbB2 tumorigenesis. Cancer Cell. 2010;17:65–76.
97. Jiang Z, Zacksenhaus E, Gallie BL, Phillips RA. The retinoblastoma gene family is differentially expressed during embryogenesis. Oncogene. 1997;14:1789–97.
98. Jirawatnotai S, Aziyu A, Osmundson EC, Moons DS, Zou XH, Kineman RD, Kiyokawa H. Cdk4 is indispensable for postnatal proliferation of the anterior pituitary. J Biol Chem. 2004;279:51100–6.
99. Jirawatnotai S, Hu YD, Livingston DM, Sicinski P. Proteomic identification of a direct role for cyclin D1 in DNA damage repair. Cancer Res. 2012;72:4289–93.
100. Jirawatnotai S, Hu YD, Michowski W, Elias JE, Becks L, Bienvenu F, Zagozdzon A, Goswami T, Wang YYE, Clark AB, et al. A function for cyclin D1 in DNA repair uncovered by protein interactome analyses in human cancers. Nature. 2011;474:230–4.
101. Jirawatnotai S, Sharma S, Michowski W, Suktitipat B, Geng Y, Quackenbush J, Elias JE, Gygi SP, Wang YYE, Sicinski P. The cyclin D1-CDK4 oncogenic interactome enables identification of potential novel oncogenes and clinical prognosis. Cell Cycle. 2014;13:2889–900.
102. Kalaszczynska I, Geng Y, Iino T, Mizuno SI, Choi Y, Kondratiuk I, Silver DP, Wolgemuth DJ, Akashi K, Sicinski P. Cyclin a is redundant in fibroblasts but essential in hematopoietic and embryonic stem cells. Cell. 2009;138:352–65.
103. Kierszenbaum AL. Cell-cycle regulation and mammalian gametogenesis: a lesson from the unexpected. Mol Reprod Dev. 2006;73:939–42.
104. Kiess M, Gill RM, Hamel PA. Expression of the positive regulator of cell cycle progression, cyclin D3, is induced during differentiation of myoblasts into quiescent myotubes. Oncogene. 1995;10:159–66.
105. Kim HA, Pomeroy SL, Whoriskey W, Pawlitzky I, Benowitz LI, Sicinski P, Stiles CD, Roberts TM. A developmentally regulated switch directs regenerative growth of Schwann cells through cyclin D1. Neuron. 2000;26:405–16.
106. Klein EA, Assoian RK. Transcriptional regulation of the cyclin D1 gene at a glance. J Cell Sci. 2008;121:3853–7.
107. Kohoutek J, Dvorak P, Hampl A. Temporal distribution of CDK4, CDK6, D-type cyclins, and p27 in developing mouse oocytes. Biol Reprod. 2004;70:139–45.
108. Kowalczyk A, Filipkowski RK, Rylski M, Wilczynski GM, Konopacki FA, Jaworski J, Ciemerych MA, Sicinski P, Kaczmarek L. The critical role of cyclin D2 in adult neurogenesis. J Cell Biol. 2004;167:209–13.
109. Kozar K, Ciemerych MA, Rebel VI, Shigematsu H, Zagozdzon A, Sicinska E, Geng Y, Yu Q, Bhattacharya S, Bronson RT, et al. Mouse development and cell proliferation in the absence of D-cyclins. Cell. 2004;118:477–91.
110. Krimpenfort P, Quon KC, Mooi WJ, Loonstra A, Berns A. Loss of p16Ink4a confers susceptibility to metastatic melanoma in mice. Nature. 2001;413:83–6.
111. Kubiak JZ, Ciemerych MA, Hupalowska A, Sikora-Polaczek M, Polanski Z. On the transition from the meiotic to mitotic cell cycle during early mouse development. Int J Dev Biol. 2008;52:201–17.
112. Kushner JA. Beta-cell growth – an unusual paradigm of organogenesis that is cyclin D2/Cdk4 dependent. Cell Cycle. 2006;5:234–7.
113. Kushner JA, Ciemerych MA, Sicinska E, Wartschow LM, Teta M, Long SY, Sicinski P, White MF. Cyclins D2 and D1 are essential for postnatal pancreatic beta-cell growth. Mol Cell Biol. 2005;25:3752–62.

114. La Baer J, Garrett MD, Stevenson LF, Slingerland JM, Sandhu C, Chou HS, Fattaey A, Harlow E. New functional activities for the p21 family of CDK inhibitors. Genes Dev. 1997;11:847–62.

115. Lam EW, Glassford J, Banerji L, Thomas NS, Sicinski P, Klaus GG. Cyclin D3 compensates for loss of cyclin D2 in mouse B-lymphocytes activated via the antigen receptor and CD40. J Biol Chem. 2000;275:3479–84.

116. Lammie GA, Fantl V, Smith R, Schuuring E, Brookes S, Michalides R, Dickson C, Arnold A, Peters G. D11S287, a putative oncogene on chromosome 11q13, is amplified and expressed in squamous cell and mammary carcinomas and linked to BCL-1. Oncogene. 1991;6:439–44.

117. Landis MW, Pawlyk BS, Li T, Sicinski P, Hinds PW. Cyclin D1-dependent kinase activity in murine development and mammary tumorigenesis. Cancer Cell. 2006;9:13–22.

118. Landis MW, Brown NE, Baker GL, Shifrin A, Das M, Geng Y, Sicinski P, Hinds PW. The LxCxE pRb interaction domain of cyclin D1 is dispensable for murine development. Cancer Res. 2007;67:7613–20.

119. Latres E, Malumbres M, Sotillo R, Martin J, Ortega S, Martin-Caballero J, Flores JM, Cordon-Cardo C, Barbacid M. Limited overlapping roles of P15(INK4b) and P18(INK4c) cell cycle inhibitors in proliferation and tumorigenesis. EMBO J. 2000;19:3496–506.

120. Lavoie JN, L'Allemain G, Brunet A, Muller R, Pouyssegur J. Cyclin D1 expression is regulated positively by the p42/p44MAPK and negatively by the p38/HOGMAPK pathway. J Biol Chem. 1996;271:20608–16.

121. LeCouter JE, Kablar B, Whyte PF, Ying C, Rudnicki MA. Strain-dependent embryonic lethality in mice lacking the retinoblastoma-related p130 gene. Development. 1998;125:4669–79.

122. LeCouter JE, Kablar B, Hardy WR, Ying C, Megeney LA, May LL, Rudnicki MA. Strain-dependent myeloid hyperplasia, growth deficiency, and accelerated cell cycle in mice lacking the Rb-related p107 gene. Mol Cell Biol. 1998;18:7455–65.

123. Lee CC, Yamamoto S, Morimura K, Wanibuchi H, Nishisaka N, Ikemoto S, Nakatani T, Wada S, Kishimoto T, Fukushima S. Significance of cyclin D1 overexpression in transitional cell carcinomas of the urinary bladder and its correlation with histopathologic features. Cancer. 1997;79:780–9.

124. Lee EY, Chang CY, Hu N, Wang YC, Lai CC, Herrup K, Lee WH, Bradley A. Mice deficient for Rb are nonviable and show defects in neurogenesis and haematopoiesis. Nature. 1992;359:288–94.

125. Lee MG, Nurse P. Complementation used to clone a human homologue of the fission yeast cell cycle control gene cdc2. Nature. 1987;327:31–5.

126. Lee MH, Williams BO, Mulligan G, Mukai S, Bronson RT, Dyson N, Harlow E, Jacks T. Targeted disruption of p107: functional overlap between p107 and Rb. Genes Dev. 1996;10:1621–32.

127. Lee Y, Dominy JE, Choi YJ, Jurczak M, Tolliday N, Camporez JP, Chim H, Lim JH, Ruan HB, Yang X, et al. Cyclin D1-Cdk4 controls glucose metabolism independently of cell cycle progression. Nature. 2014;510:547–51.

128. Lew DJ, Dulic V, Reed SI. Isolation of three novel human cyclins by rescue of G1 cyclin (Cln) function in yeast. Cell. 1991;66:1197–206.

129. Lewandoski M. Conditional control of gene expression in the mouse. Nat Rev Genet. 2001;2:743–55.

130. Li Z, Wang C, Jiao X, Katiyar S, Casimiro MC, Prendergast GC, Powell MJ, Pestell RG. Alternate cyclin d1 mRNA splicing modulates p27(KIP1) binding and cell migration. J Biol Chem. 2008;283:7007–15.

131. Li ZP, Jiao XM, Wang CG, Shirley LA, Elsaleh H, Dahl O, Wang M, Soutoglou E, Knudsen ES, Pestell RG. Alternative cyclin D1 splice forms differentially regulate the DNA damage response. Cancer Res. 2010;70:8802–11.

132. Lin DI, Lessie MD, Gladden AB, Bassing CH, Wagner KU, Diehl JA. Disruption of cyclin D1 nuclear export and proteolysis accelerates mammary carcinogenesis. Oncogene. 2008;27:1231–42.

133. Liu F. Smad3 phosphorylation by cyclin-dependent kinases. Cytokine Growth Factor Rev. 2006;17(1–2):9–17.
134. Liu L, Michowski W, Inuzuka H, Shimizu K, Nihira NT, Chick JM, Li N, Geng Y, Meng AY, Ordureau A, et al. G1 cyclins link proliferation, pluripotency and differentiation of embryonic stem cells. Nat Cell Biol. 2017;19:177–88.
135. Liu WD, Wang HW, Muguira M, Breslin MB, Lan MS. INSM1 functions as a transcriptional repressor of the neuroD/beta 2 gene through the recruitment of cyclin D1 and histone deacetylases. Biochem J. 2006;397:169–77.
136. Lucas JJ, Szepesi A, Modiano JF, Domenico J, Gelfand EW. Regulation of synthesis and activity of the plstire protein (cyclin-dependent kinase-6 (Cdk6)), a major cyclin-D-associated Cdk4 homolog in normal human T-lymphocytes. J Immunol. 1995;154:6275–84.
137. Ma CY, Papermaster D, Cepko CL. A unique pattern of photoreceptor degeneration in cyclin D1 mutant mice. Proc Natl Acad Sci U S A. 1998;95:9938–43.
138. Maandag EC, van der Valk M, Vlaar M, Feltkamp C, O'Brien J, van Roon M, van der Lugt N, Berns A, te Riele H. Developmental rescue of an embryonic-lethal mutation in the retinoblastoma gene in chimeric mice. EMBO J. 1994;13:4260–8.
139. MacPherson D, Sage J, Crowley D, Trumpp A, Bronson RT, Jacks T. Conditional mutation of Rb causes cell cycle defects without apoptosis in the central nervous system. Mol Cell Biol. 2003;23:1044–53.
140. Maeda K, Chung Y, Kang S, Ogawa M, Onoda N, Nishiguchi Y, Ikehara T, Nakata B, Okuno M, Sowa M. Cyclin D1 overexpression and prognosis in colorectal adenocarcinoma. Oncology. 1998;55:145–51.
141. Mak TW. Gene targeting in embryonic stem cells scores a knockout in Stockholm. Cell. 2007;131:1027–31.
142. Malumbres M. Cyclin-dependent kinases. Genome Biol. 2014;15:122.
143. Malumbres M, Barbacid M. Cell cycle, CDKs and cancer: a changing paradigm. Nat Rev Cancer. 2009;9:153–66.
144. Malumbres M, Sotillo R, Santamaria D, Galan J, Cerezo A, Ortega S, Dubus P, Barbacid M. Mammalian cells cycle without the D-type Cyclin-dependent kinases Cdk4 and Cdk6. Cell. 2004;118:493–504.
145. Mariappan I, Parnaik VK. Sequestration of pRb by cyclin D3 causes intranuclear reorganization of lamin A/C during muscle cell differentiation. Mol Biol Cell. 2005;16:1948–60.
146. Martin GR. Isolation of a pluripotent cell line from early mouse embryos cultured in medium conditioned by teratocarcinoma stem cells. Proc Natl Acad Sci U S A. 1981;78:7634–8.
147. Martin J, Hunt SL, Dubus P, Sotillo R, Nehme-Pelluard F, Magnuson MA, Parlow AF, Malumbres M, Ortega S, Barbacid M. Genetic rescue of Cdk4 null mice restores pancreatic beta-cell proliferation but not homeostatic cell number. Oncogene. 2003;22:5261–9.
148. Masui Y. From oocyte maturation to the in vitro cell cycle: the history of discoveries of Maturation-Promoting Factor (MPF) and Cytostatic Factor (CSF). Differentiation. 2001;69:1–17.
149. Masui Y, Markert CL. Cytoplasmic control of nuclear behavior during meiotic maturation of frog oocytes. J Exp Zool. 1971;177:129–45.
150. Matараza JM, Tumang JR, Gumina MR, Gurdak SM, Rothstein TL, Chiles TC. Disruption of cyclin D3 blocks proliferation of normal B-1a cells, but loss of cyclin D3 is compensated by cyclin D2 in cyclin D3-deficient mice. J Immunol. 2006;177:787–95.
151. Mate JL, Ariza A, Aracil C, Lopez D, Isamat M, Perez-Piteira J, Navas-Palacios JJ. Cyclin D1 overexpression in non-small cell lung carcinoma: correlation with Ki67 labelling index and poor cytoplasmic differentiation. J Pathol. 1996;180:395–9.
152. Matsuoka S, Yamaguchi M, Matsukage A. D-type Cyclin-binding regions of proliferating cell nuclear antigen. J Biol Chem. 1994;269:11030–6.
153. Matsushime H, Roussel MF, Ashmun RA, Sherr CJ. Colony-stimulating factor 1 regulates novel cyclins during the G1 phase of the cell cycle. Cell. 1991;65:701–13.
154. Matsushime H, Quelle DE, Shurtleff SA, Shibuya M, Sherr CJ, Kato JY. D-type cyclin-dependent kinase activity in mammalian cells. Mol Cell Biol. 1994;14:2066–76.

155. Matsushime H, Ewen ME, Strom DK, Kato JY, Hanks SK, Roussel MF, Sherr CJ. Identification and properties of an atypical catalytic subunit (p34PSK-J3/cdk4) for mammalian D type G1 cyclins. Cell. 1992;71:323–34.
156. Matsuura I, Denissova NG, Wang G, He D, Long J, Liu F. Cyclin-dependent kinases regulate the antiproliferative function of Smads. Nature. 2004;430:226–31.
157. Mettus RV, Rane SG. Characterization of the abnormal pancreatic development, reduced growth and infertility in Cdk4 mutant mice. Oncogene. 2003;22:8413–21.
158. Meyerson M, Harlow E. Identification of G1 kinase activity for cdk6, a novel cyclin D partner. Mol Cell Biol. 1994;14:2077–86.
159. Meyerson M, Enders GH, Wu CL, Su LK, Gorka C, Nelson C, Harlow E, Tsai LH. A family of human cdc2-related protein kinases. EMBO J. 1992;11:2909–17.
160. Miliani De Marval PL, Macias E, Rounbehler R, Sicinski P, Kiyokawa H, Johnson DG, Conti CJ, Rodriguez-Puebla ML. Lack of cyclin-dependent kinase 4 inhibits c-myc tumorigenic activities in epithelial tissues. Mol Cell Biol. 2004;24:7538–47.
161. Momčilović O, Navara C, Schatten G. Cell cycle adaptations and maintenance of genomic integrity in embryonic stem cells and induced pluripotent stem cells. In: Kubiak JZ, editor. Cell cycle in development. Berlin/Heidelberg: Springer-Verlag GmbH; 2011.
162. Moons DS, Jirawatnotai S, Parlow AF, Gibori G, Kineman RD, Kiyokawa H. Pituitary hypoplasia and lactotroph dysfunction in mice deficient for cyclin-dependent kinase-4. Endocrinology. 2002;143:3001–8.
163. Moons DS, Jirawatnotai S, Tsutsui T, Franks R, Parlow AF, Hales DB, Gibori G, Fazleabas AT, Kiyokawa H. Intact follicular maturation and defective luteal function in mice deficient ford cyclin-dependent kinase-4. Endocrinology. 2002;143:647–54.
164. Moore GD, Ayabe T, Kopf GS, Schultz RM. Temporal patterns of gene expression of G1-S cyclins and cdks during the first and second mitotic cell cycles in mouse embryos. Mol Reprod Dev. 1996;45:264–75.
165. Motokura T, Keyomarsi K, Kronenberg HM, Arnold A. Cloning and characterization of human cyclin D3, a cDNA closely related in sequence to the PRAD1/cyclin D1 proto-oncogene. J Biol Chem. 1992;267:20412–5.
166. Motokura T, Bloom T, Kim HG, Juppner H, Ruderman JV, Kronenberg HM, Arnold A. A novel cyclin encoded by a bcl1-linked candidate oncogene. Nature. 1991;350:512–5.
167. Musgrove EA. Cyclins: roles in mitogenic signaling and oncogenic transformation. Growth Factors. 2006;24:13–9.
168. Musgrove EA, Caldon CE, Barraclough J, Stone A, Sutherland RL. Cyclin D as a therapeutic target in cancer. Nat Rev Cancer. 2011;11:558–72.
169. Neuman E, Ladha MH, Lin N, Upton TM, Miller SJ, Di Renzo J, Pestell RG, Hinds PW, Dowdy SF, Brown M, et al. Cyclin D1 stimulation of estrogen receptor transcriptional activity independent of cdk4. Mol Cell Biol. 1997;17:5338–47.
170. Neumeister P, Pixley FJ, Xiong Y, Xie HF, Wu KM, Ashton A, Cammer M, Chan A, Symons M, Stanley ER, et al. Cyclin D1 governs adhesion and motility of macrophages. Mol Biol Cell. 2003;14:2005–15.
171. Nobs L, Nestel S, Kulik A, Nitsch C, Atanasoski S. Cyclin D1 is required for proliferation of olig2-expressing progenitor cells in the injured cerebral cortex. Glia. 2013;61:1443–55.
172. Odajima J, Saini S, Jung P, Ndassa-Colday Y, Ficaro S, Geng Y, Marco E, Michowski W, Wang YE, DeCaprio JA, et al. Proteomic landscape of tissue-specific cyclin E functions in vivo. PLoS Genet. 2016;12:e1006429.
173. Ogasawara T, Chikuda H, Ohba S, Chikazu D, Katagiri M, Yano F, Nakamura K, Chung U, Hoshi K, Takato T, et al. Functional switching of Runx2 by Cdk6 and Cdk4 in regulation of osteoblast differentiation. J Bone Miner Res. 2005;20:S5.
174. Opitz OG, Rustgi AK. Interaction between Sp1 and cell cycle regulatory proteins is important in transactivation of a differentiation-related gene. Cancer Res. 2000;60:2825–30.
175. Palmero I, Holder A, Sinclair AJ, Dickson C, Peters G. Cyclins D1 and D2 are differentially expressed in human B-lymphoid cell lines. Oncogene. 1993;8:1049–54.

176. Pauling JK, Christensen AG, Batra R, Alcaraz N, Barbosa E, Larsen MR, Beck HC, Leth-Larsen R, Azevedo V, Ditzel HJ, et al. Elucidation of epithelial mesenchymal transition-related pathways in a triple-negative breast cancer cell line model by multi-omics interactome analysis. Integr Biol-Uk. 2014;6:1058–68.

177. Perez-Roger I, Kim SH, Griffiths B, Sewing A, Land H. Cyclins D1 and D2 mediate myc-induced proliferation via sequestration of p27(Kip1) and p21(Cip1). EMBO J. 1999;18:5310–20.

178. Peters G, Fantl V, Smith R, Brookes S, Dickson C. Chromosome 11q13 markers and D-type cyclins in breast cancer. Breast Cancer Res Treat. 1995;33:125–35.

179. Pirkmaier A, Dow R, Ganiatsas S, Waring P, Warren K, Thompson A, Hendley J, Germain D. Alternative mammary oncogenic pathways are induced by D-type cyclins; MMTV-cyclin D3 transgenic mice develop squamous cell carcinoma. Oncogene. 2003;22:4425–33.

180. Quelle DE, Ashmun RA, Hannon GJ, Rehberger PA, Trono D, Richter KH, Walker C, Beach D, Sherr CJ, Serrano M. Cloning and characterization of murine p16INK4a and p15INK4b genes. Oncogene. 1995;11:635–45.

181. Rane SG, Cosenza SC, Mettus RV, Reddy EP. Germ line transmission of the Cdk4(R24C) mutation facilitates tumorigenesis and escape from cellular senescence. Mol Cell Biol. 2002;22:644–56.

182. Rane SG, Dubus P, Mettus RV, Galbreath EJ, Boden G, Reddy EP, Barbacid M. Loss of Cdk4 expression causes insulin-deficient diabetes and Cdk4 activation results in beta-islet cell hyperplasia. Nat Genet. 1999;22:44–52.

183. Rao PN, Wilson B, Puck TT. Premature chromosome condensation and cell cycle analysis. J Cell Physiol. 1977;91:131–41.

184. Rao SS, Chu C, Kohtz DS. Ectopic expression of cyclin D1 prevents activation of gene transcription by myogenic basic helix-loop-helix regulators. Mol Cell Biol. 1994;14:5259–67.

185. Ratineau C, Petry MW, Mutoh H, Leiter AB. Cyclin D1 represses the basic helix-loop-helix transcription factor, BETA2/NeuroD. J Biol Chem. 2002;277:8847–53.

186. Ravnik SE, Rhee K, Wolgemuth DJ. Distinct patterns of expression of the D-type cyclins during testicular development in the mouse. Dev Genet. 1995;16:171–8.

187. Reddy HK, Grana X, Dhanasekaran DN, Litvin J, Reddy EP. Requirement of Cdk4 for v-Ha-ras-induced breast tumorigenesis and activation of the v-ras-induced senescence program by the R24C mutation. Genes Cancer. 2010;1:69–80.

188. Reddy HK, Mettus RV, Rane SG, Grana X, Litvin J, Reddy EP. Cyclin-dependent kinase 4 expression is essential for neu-induced breast tumorigenesis. Cancer Res. 2005;65:10174–8.

189. Reutens AT, Fu MF, Wang CG, Albanese C, McPhaul MJ, Sun ZJ, Balk SP, Janne OA, Palvimo JJ, Pestell RG. Cyclin D1 binds the androgen receptor and regulates hormone-dependent signaling in a p300/CBP-associated factor (P/CAF)-dependent manner. Mol Endocrinol. 2001;15:797–811.

190. Robles AI, Rodriguez-Puebla ML, Glick AB, Trempus C, Hansen L, Sicinski P, Tennant RW, Weinberg RA, Yuspa SH, Conti CJ. Reduced skin tumor development in cyclin D1-deficient mice highlights the oncogenic ras pathway in vivo. Genes Dev. 1998;12:2469–74.

191. Rodriguez-Puebla ML, Miliani de Marval PL, LaCava M, Moons DS, Kiyokawa H, Conti CJ. Cdk4 deficiency inhibits skin tumor development but does not affect normal keratinocyte proliferation. Am J Pathol. 2002;161:405–11.

192. Rosenthal ET, Hunt T, Ruderman JV. Selective translation of messenger-Rna controls the pattern of protein-synthesis during early development of the surf clam, spisula-solidissima. Cell. 1980;20:487–94.

193. Rossant J, McMahon A. Creating mouse mutants-a meeting review on conditional mouse genetics. Genes Dev. 1999;13:142–5.

194. Russell A, Thompson MA, Hendley J, Trute L, Armes J, Germain D. Cyclin D1 and D3 associate with the SCF complex and are coordinately elevated in breast cancer. Oncogene. 1999;18:1983–91.

195. Saavedra-Avila NA, Sengupta U, Sanchez B, Sala E, Haba L, Stratmann T, Verdaguer J, Mauricio D, Mezquita B, Ropero AB, et al. Cyclin D3 promotes pancreatic beta-cell fitness and viability in a cell cycle-independent manner and is targeted in autoimmune diabetes. Proc Natl Acad Sci U S A. 2014;111:E3405–14.
196. Sage J, Mulligan GJ, Attardi LD, Miller A, Chen S, Williams B, Theodorou E, Jacks T. Targeted disruption of the three Rb-related genes leads to loss of G(1) control and immortalization. Genes Dev. 2000;14:3037–50.
197. Said TK, Conneely OM, Medina D, O'Malley BW, Lydon JP. Progesterone, in addition to estrogen, induces cyclin D1 expression in the murine mammary epithelial cell, in vivo. Endocrinology. 1997;138:3933–9.
198. Sallinen SL, Sallinen PK, Kononen JT, Syrjakoski KM, Nupponen NN, Rantala IS, Helen PT, Helin HJ, Haapasalo HK. Cyclin D1 expression in astrocytomas is associated with cell proliferation activity and patient prognosis. J Pathol. 1999;188:289–93.
199. Sankaran VG, Ludwig LS, Sicinska E, Xu J, Bauer DE, Eng JC, Patterson HC, Metcalf RA, Natkunam Y, Orkin SH, et al. Cyclin D3 coordinates the cell cycle during differentiation to regulate erythrocyte size and number. Genes Dev. 2012;26:2075–87.
200. Santamaria D, Barriere C, Cerqueira A, Hunt S, Tardy C, Newton K, Caceres JF, Dubus P, Malumbres M, Barbacid M. Cdk1 is sufficient to drive the mammalian cell cycle. Nature. 2007;448:811–5.
201. Savatier P, Lapillonne H, van Grunsven LA, Rudkin BB, Samarut J. Withdrawal of differentiation inhibitory activity/leukemia inhibitory factor up-regulates D-type cyclins and cyclin-dependent kinase inhibitors in mouse embryonic stem cells. Oncogene. 1996;12:309–22.
202. Savatier P, Lapillonne H, Jirmanova L, Vitelli L, Samarut J. Analysis of the cell cycle in mouse embryonic stem cells. Methods Mol Biol. 2002;185:27–33.
203. Scheicher R, Hoelbl-Kovacic A, Bellutti F, Tigan AS, Prchal-Murphy M, Heller G, Schneckenleithner C, Salazar-Roa M, Zochbauer-Muller S, Zuber J, et al. CDK6 as a key regulator of hematopoietic and leukemic stem cell activation. Blood. 2015;125:90–101.
204. Serrano M, Hannon GJ, Beach D. A new regulatory motif in cell-cycle control causing specific inhibition of cyclin D/CDK4. Nature. 1993;366:704–7.
205. Serrano M, Lee H, Chin L, Cordon-Cardo C, Beach D, De Pinho RA. Role of the INK4a locus in tumor suppression and cell mortality. Cell. 1996;85:27–37.
206. Sharpless NE, Ramsey MR, Balasubramanian P, Castrillon DH, DePinho RA. The differential impact of p16(INK4a) or p19(ARF) deficiency on cell growth and tumorigenesis. Oncogene. 2004;23:379–85.
207. Sharpless NE, Bardeesy N, Lee KH, Carrasco D, Castrillon DH, Aguirre AJ, Wu EA, Horner JW, DePinho RA. Loss of p16Ink4a with retention of p19Arf predisposes mice to tumorigenesis. Nature. 2001;413:86–91.
208. Shen T, Huang SL. The role of Cdc25A in the regulation of cell proliferation and apoptosis. Anti-Cancer Agent Me. 2012;12:631–9.
209. Sherr CJ, Roberts JM. CDK inhibitors: positive and negative regulators of G1-phase progression. Genes Dev. 1999;13:1501–12.
210. Sherr CJ, Roberts JM. Living with or without cyclins and cyclin-dependent kinases. Genes Dev. 2004;18:2699–711.
211. Sicinska E, Aifantis I, Le Cam L, Swat W, Borowski C, Yu Q, Ferrando AA, Levin SD, Geng Y, von Boehmer H, et al. Requirement for cyclin D3 in lymphocyte development and T cell leukemias. Cancer Cell. 2003;4:451–61.
212. Sicinska E, Lee YM, Gits J, Shigematsu H, Yu Q, Rebel VI, Geng Y, Marshall CJ, Akashi K, Dorfman DM, et al. Essential role for cyclin D3 in granulocyte colony-stimulating factor-driven expansion of neutrophil granulocytes. Mol Cell Biol. 2006;26:8052–60.
213. Sicinski P, Donaher JL, Parker SB, Li T, Fazeli A, Gardner H, Haslam SZ, Bronson RT, Elledge SJ, Weinberg RA. Cyclin D1 provides a link between development and oncogenesis in the retina and breast. Cell. 1995;82:621–30.
214. Sicinski P, Donaher JL, Geng Y, Parker SB, Gardner H, Park MY, Robker RL, Richards JS, McGinnis LK, Biggers JD, et al. Cyclin D2 is an FSH-responsive gene involved in gonadal cell proliferation and oncogenesis. Nature. 1996;384:470–4.

215. Simanis V, Nurse P. The cell cycle control gene cdc2+ of fission yeast encodes a protein kinase potentially regulated by phosphorylation. Cell. 1986;45:261–8.
216. Simmons J. Virchow and the cell doctrine. In: Scientific 100 – a ranking of most influential scientists, past and present. Secaucus: Carol Publishing Group; 1996. p. 88–92.
217. Skapek SX, Rhee J, Spicer DB, Lassar AB. Inhibition of myogenic differentiation in proliferating myoblasts by cyclin D1-dependent kinase. Science. 1995;267:1022–4.
218. Skapek SX, Rhee J, Kim PS, Novitch BG, Lassar AB. Cyclin-mediated inhibition of muscle gene expression via a mechanism that is independent of pRB hyperphosphorylation. Mol Cell Biol. 1996;16:7043–53.
219. Smith LD, Ecker RE. The interaction of steroids with Rana pipiens oocytes in the induction of maturation. Dev Biol. 1971;25:232–47.
220. Solvason N, Wu WW, Parry D, Mahony D, Lam EW, Glassford J, Klaus GG, Sicinski P, Weinberg R, Liu YJ, et al. Cyclin D2 is essential for BCR-mediated proliferation and CD5 B cell development. Int Immunol. 2000;12:631–8.
221. Sunkara PS, Al-Bader AA, Riker MA, Rao PN. Induction of prematurely condensed chromosomes by mitoplasts. Cell Biol Int Rep. 1980;4:1025–9.
222. Swaminathan G, Varghese B, Fuchs SY. Regulation of prolactin receptor levels and activity in breast cancer. J Mammary Gland Biol Neoplasia. 2008;13:81–91.
223. Sweeney KJ, Swarbrick A, Sutherland RL, Musgrove EA. Lack of relationship between CDK activity and G1 cyclin expression in breast cancer cells. Oncogene. 1998;16:2865–78.
224. Swenson KI, Farrell KM, Ruderman JV. The clam embryo protein cyclin a induces entry into M phase and the resumption of meiosis in Xenopus oocytes. Cell. 1986;47:861–70.
225. Takaki T, Fukasawa K, Suzuki-Takahashi I, Semba K, Kitagawa M, Taya Y, Hirai H. Preferences for phosphorylation sites in the retinoblastoma protein of D-type cyclindependent kinases, Cdk4 and Cdk6, in vitro. J Biochem (Tokyo). 2005;137:381–6.
226. Tam SW, Theodoras AM, Shay JW, Draetta GF, Pagano M. Differential expression and regulation of Cyclin D1 protein in normal and tumor human cells: association with Cdk4 is required for Cyclin D1 function in G1 progression. Oncogene. 1994;9:2663–74.
227. Taneja P, Frazier DP, Kendig RD, Maglic D, Sugiyama T, Kai F, Taneja NK, Inoue K. MMTV mouse models and the diagnostic values of MMTV-like sequences in human breast cancer. Expert Rev Mol Diagn. 2009;9:423–40.
228. Tapias A, Ciudad CJ, Roninson IB, Noe V. Regulation of Sp1 by cell cycle related proteins. Cell Cycle. 2008;7:2856–67.
229. Tarkowski AK. Mouse chimaeras developed from fused eggs. Nature. 1961;190:857–60.
230. Tarkowski AK. Mouse chimaeras revisited: recollections and reflections. Int J Dev Biol. 1998;42:903–8.
231. Tsutsui T, Hesabi B, Moons DS, Pandolfi PP, Hansel KS, Koff A, Kiyokawa H. Targeted disruption of CDK4 delays cell cycle entry with enhanced p27(Kip1) activity. Mol Cell Biol. 1999;19:7011–9.
232. Tworoger SS, Hankinson SE. Prolactin and breast cancer etiology: an epidemiologic perspective. J Mammary Gland Biol Neoplasia. 2008;13:41–53.
233. van der Kuip H, Carius B, Haque SJ, Williams BRG, Huber C, Fischer T. The DNA-binding subunit p140 of replication factor C is upregulated in cycling cells and associates with G(1) phase cell cycle regulatory proteins. J Mol Med-JMM. 1999;77:386–92.
234. Wang CG, Fan SJ, Li ZP, Fu MF, Rao M, Ma YX, Lisanti MP, Albanese C, Katzenellenbogen BS, Kushner PJ, et al. Cyclin D1 antagonizes BRCA1 repression of estrogen receptor alpha activity. Cancer Res. 2005;65:6557–67.
235. Wang CG, Pattabiraman N, Zhou JN, Fu MF, Sakamaki T, Albanese C, Li ZP, Wu KM, Hulit J, Neumeister P, et al. Cyclin D1 repression of peroxisome proliferator-activated receptor gamma expression and transactivation. Mol Cell Biol. 2003;23:6159–73.
236. Wang TC, Cardiff RD, Zukerberg L, Lees E, Arnold A, Schmidt EV. Mammary hyperplasia and carcinoma in MMTV-cyclin D1 transgenic mice. Nature. 1994;369:669–71.
237. Wenzel PL, Wu L, de Bruin A, Chong JL, Chen WY, Dureska G, Sites E, Pan T, Sharma A, Huang K, et al. Rb is critical in a mammalian tissue stem cell population. Genes Dev. 2007;21:85–97.

238. White J, Stead E, Faast R, Conn S, Cartwright P, Dalton S. Developmental activation of the Rb-E2F pathway and establishment of cell cycle-regulated cyclin-dependent kinase activity during embryonic stem cell differentiation. Mol Biol Cell. 2005;16:2018–27.

239. Wianny F, Real FX, Mummery CL, Van Rooijen M, Lahti J, Samarut J, Savatier P. G1-phase regulators, cyclin D1, cyclin D2, and cyclin D3: up-regulation at gastrulation and dynamic expression during neurulation. Dev Dyn. 1998;212:49–62.

240. Williams BO, Schmitt EM, Remington L, Bronson RT, Albert DM, Weinberg RA, Jacks T. Extensive contribution of Rb-deficient cells to adult chimeric mice with limited histopathological consequences. EMBO J. 1994;13:4251–9.

241. Won KA, Xiong Y, Beach D, Gilman MZ. Growth-regulated expression of D-type cyclin genes in human diploid fibroblasts. Proc Natl Acad Sci U S A. 1992;89:9910–4.

242. Wu L, de Bruin A, Saavedra HI, Starovic M, Trimboli A, Yang Y, Opavska J, Wilson P, Thompson JC, Ostrowski MC, et al. Extra-embryonic function of Rb is essential for embryonic development and viability. Nature. 2003;421:942–7.

243. Xiong Y, Zhang H, Beach D. D type cyclins associate with multiple protein kinases and the DNA replication and repair factor PCNA. Cell. 1992;71:505–14.

244. Xiong Y, Connolly T, Futcher B, Beach D. Human D-type cyclin. Cell. 1991;65:691–9.

245. Xiong Y, Menninger J, Beach D, Ward DC. Molecular cloning and chromosomal mapping of CCND genes encoding human D-type cyclins. Genomics. 1992;13:575–84.

246. Yang C, Chen L, Li C, Lynch MC, Brisken C, Schmidt EV. Cyclin D1 enhances the response to estrogen and progesterone by regulating progesterone receptor expression. Mol Cell Biol. 2010;30:3111–25.

247. Yang C, Ionescu-Tiba V, Burns K, Gadd M, Zukerberg L, Louis DN, Sgroi D, Schmidt EV. The role of the cyclin D1-dependent kinases in ErbB2-mediated breast cancer. Am J Pathol. 2004;164:1031–8.

248. Yang R, Bie W, Haegebarth A, Tyner AL. Differential regulation of D-type cyclins in the mouse intestine. Cell Cycle. 2006;5:180–3.

249. Yu Q, Geng Y, Sicinski P. Specific protection against breast cancers by cyclin D1 ablation. Nature. 2001;411:1017–21.

250. Yu Q, Ciemerych MA, Sicinski P. Ras and Myc can drive oncogenic cell proliferation through individual D-cyclins. Oncogene. 2005;24:7114–9.

251. Yu Q, Sicinska E, Geng Y, Ahnstrom M, Zagozdzon A, Kong Y, Gardner H, Kiyokawa H, Harris LN, Stal O, et al. Requirement for CDK4 kinase function in breast cancer. Cancer Cell. 2006;9:23–32.

252. Zacksenhaus E, Jiang Z, Chung D, Marth JD, Phillips RA, Gallie BL. pRb controls proliferation, differentiation, and death of skeletal muscle cells and other lineages during embryogenesis. Genes Dev. 1996;10:3051–64.

253. Zarkowska T, Mittnacht S. Differential phosphorylation of the retinoblastoma protein by G1/S cyclin-dependent kinases. J Biol Chem. 1997;272:12738–46.

254. Zhang Q, Wang XY, Wolgemuth DJ. Developmentally regulated expression of cyclin D3 and its potential in vivo interacting proteins during murine gametogenesis. Endocrinology. 1999;140:2790–800.

255. Zhang Q, Sakamoto K, Liu C, Triplett AA, Lin WC, Rui H, Wagner KU. Cyclin D3 compensates for the loss of cyclin D1 during ErbB2-induced mammary tumor initiation and progression. Cancer Res. 2011;71:7513–24.

256. Zheng B, Sage M, Sheppeard EA, Jurecic V, Bradley A. Engineering mouse chromosomes with Cre-loxP: range, efficiency, and somatic applications. Mol Cell Biol. 2000;20:648–55.

257. Zindy F, Quelle DE, Roussel MF, Sherr CJ. Expression of the p16INK4a tumor suppressor versus other INK4 family members during mouse development and aging. Oncogene. 1997;15:203–11.

258. Zindy F, van Deursen J, Grosveld G, Sherr CJ, Roussel MF. INK4d-deficient mice are fertile despite testicular atrophy. Mol Cell Biol. 2000;20:372–8.

259. Zindy F, Soares H, Herzog KH, Morgan J, Sherr CJ, Roussel MF. Expression of INK4 inhibitors of cyclin D-dependent kinases during mouse brain development. Cell Growth Differ. 1997;8:1139–50.
260. Zindy F, den Besten W, Chen B, Rehg JE, Latres E, Barbacid M, Pollard JW, Sherr CJ, Cohen PE, Roussel MF. Control of spermatogenesis in mice by the cyclin D-dependent kinase inhibitors p18(Ink4c) and p19(Ink4d). Mol Cell Biol. 2001;21:3244–55.
261. Zou X, Ray D, Aziyu A, Christov K, Boiko AD, Gudkov AV, Kiyokawa H. Cdk4 disruption renders primary mouse cells resistant to oncogenic transformation, leading to Arf/p53-independent senescence. Genes Dev. 2002;16:2923–34.
262. Zukerberg LR, Yang WI, Gadd M, Thor AD, Koerner FC, Schmidt EV, Arnold A. Cyclin D1 (PRAD1) protein expression in breast cancer: approximately one-third of infiltrating mammary carcinomas show overexpression of the cyclin D1 oncogene. Mod Pathol. 1995;8:560–7.
263. Zwijsen RM, Wientjens E, Klompmaker R, van der Sman J, Bernards R, Michalides RJ. CDK-independent activation of estrogen receptor by cyclin D1. Cell. 1997;88:405–15.

Chapter 3
D-Type Cyclins and Gene Transcription

Gabriele Di Sante, Mathew C. Casimiro, Zhiping Li, Adam Ertel,
Peter Tompa, and Richard G. Pestell

Abstract D-type cyclins contribute the regulatory subunits to the holoenzymes
that phosphorylate distinct substrates and regulate diverse biological processes,
including cellular proliferation and differentiation. A growing body of evidence
has demonstrated that the D-type cyclins are located in distinct subcellular pools
with distinct functions. These subcellular locations include the cell membrane, the
cytoplasm, the nuclear lamina, the nucleus, and DNA binding sites. The distribution of D-type cyclins in each one of these compartments is regulated by distinct
signaling pathways and contributes to a vast array of biological processes.
Importantly, D-type cyclins can also conduct transcriptional functions. Initially
shown to regulate the activity of transcription factors in gene reporter assays, subsequent studies have shown that D-type cyclins can affect gene expression through
the regulation of coactivators, the modulation of transcription factor binding in the
context of chromatin, and their ability to serve as molecular scaffolds that facilitate

Gabriele Di Sante and Mathew C. Casimiro contributed equally to this work.

G. Di Sante • M.C. Casimiro
Pennsylvania Cancer and Regenerative Medicine Research Center,
Baruch S. Blumberg Institute, Pennsylvania Biotechnology Center,
100 East Lancaster Avenue, Suite, 222, Wynnewood, PA 19096, USA

Z. Li • A. Ertel
Department of Cancer Biology, Thomas Jefferson University,
233 South 10th Street, Philadelphia, PA 19107, USA

P. Tompa
VIB center for Structural Biology (CSB), Brussels, Belgium

Structural Biology Brussels (SBB), Vrije Universiteit Brussel (VUB), Brussels, Belgium

Institute of Enzymology, Research Centre for Natural Sciences of the Hungarian
Academy of Sciences, Budapest, Hungary

R.G. Pestell (✉)
Pennsylvania Cancer and Regenerative Medicine Research Center,
Baruch S. Blumberg Institute, Pennsylvania Biotechnology Center,
100 East Lancaster Avenue, Suite, 222, Wynnewood, PA 19096, USA

Lee Kong Chian School of Medicine, Nanyang Technological University,
639798, Singapore
e-mail: Richard.Pestell@gmail.com

© Springer International Publishing AG 2018
P.W. Hinds, N.E. Brown (eds.), *D-type Cyclins and Cancer*,
Current Cancer Research, DOI 10.1007/978-3-319-64451-6_3

the interactions between chromatin-modifying enzymes (histone and DNA methylases [Suv39, HP1α, G9a, DNMT1, PRMT1], histone acetylases [SRC1, p300, P/CAF], and histone deacetylases [HDAC1,3]). Genome-wide studies have shown that cyclin D1 associates with the regulatory regions of more than 2840 genes. The importance of the transcriptional functions of D-type cyclins has been demonstrated in vivo, as hormone signaling in prostate and mammary gland tissues is critically dependent upon the presence of cyclin D1. Furthermore, it is the cyclin D1-regulated gene expression signature, not the abundance of the protein, which strongly predicts poor outcome in prostate cancer. Together, these findings are consistent with the evolving realization that the D-type cyclins play an important biological role in governing gene transcription.

Keywords Cyclins • Cdk • transcription factor • Chromatin • Histone acetylation • Estrogen signaling • Androgen signaling

3.1   Introduction

The mechanisms through which cell cycle control proteins regulate gene expression, both by altering the levels of transcription factors and by regulating the basal transcriptional apparatus, have been under scrutiny for more than two decades [1–3]. Among the first principles outlined, it was proposed that, in addition to governing kinases that phosphorylate cell cycle control proteins, some transcription factors were also targets of cell cycle control kinases. The evidence thus supported a model in which cyclin-cyclin-dependent kinase (cyclin-Cdk) complexes could regulate transcription factor (TF) activity through phosphorylation that altered TF binding to DNA, altered the recruitment of cofactors, or otherwise led to alterations in their subcellular localization [1]. In part, these functions were thought to involve interactions with histone acetylases [4, 5]. Since these early beginnings, a substantial expansion in our understanding of the mechanisms linking cell cycle proteins and transcription has occurred. Using new experimental tools, the biological significance of the transcriptional functions of the cyclin-Cdk complexes has become clearer.

It is now known that D-type cyclins regulate TF activity through three well-characterized activities [6]:

A. Regulation of TF recruitment to TF binding sites [7, 8].
B. Histone modifications, in particular acetylation/deacetylation of H3K9 and H3K27. Transcriptionally active chromatin correlates with deacetylation of histone H3K9 [9, 10].
C. Recruitment of chromatin-modifying enzymes (Suv39, HP1α, p300/CBP, HDAC1/3, P/CAF, G9a, DNMT1, PRMT5) at TF sites [7, 11–13].

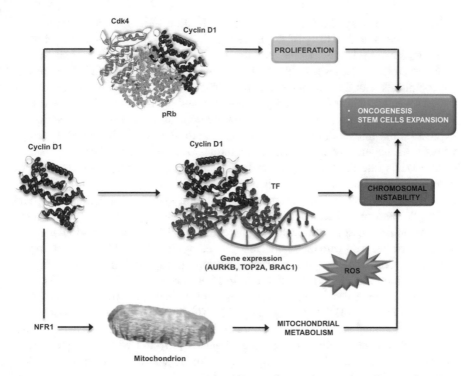

Fig. 3.1 The dual role of cyclin D1 in promoting oncogenesis and the expansion of stem cells. Cyclin D1 (*red*) promotes cell proliferation and stem cell expansion through distinct mechanisms, both kinase-dependent and kinase-independent. Cyclin D1 forms a complex with cyclin-dependent kinase 4 (Cdk4, *yellow*), which phosphorylates the retinoblastoma protein (*pRb*), leading to cell cycle progression and proliferation. However, cyclin D1, by binding specific transcription factors (*blue*), is also able to occupy the promoter regions of genes involved in the chromosomal instability, enhancing their expression. Moreover, chromosomal stability is also regulated by cyclin D1 by regulating the production of reactive oxygen species (*ROS*). Indeed, cyclin D1 has a role in mitochondrial metabolism through the NRF1-mediated pathway

Consistent with all these functions, for example, cyclin D1 can bind gene regulatory regions [1, 6, 7, 11, 12, 14, 15], a finding recently confirmed by ChIP of local promoter regions [16] and ChIP-Seq analysis [17]. This review outlines the conceptual progresses in our understanding of the biological significance and mechanisms by which the D-type cyclins govern gene transcription (Fig. 3.1).

3.2   The History of D-Type Cyclins

Cyclin-dependent kinases (Cdks) are a family of serine/threonine kinases in charge of controlling progression through the cell cycle [1, 18–22]. Regulatory subunits, known as cyclins, form complexes with Cdks and thus phosphorylate specific proteins at different phases of the cell cycle [1, 21–23]. The discovery of cyclins in the

1980s, as proteins synthesized during the fertilization of marine invertebrate eggs [24], and the later discovery of the human [25, 26] and murine D-type cyclins in 1991 were pivotal landmarks for the later harnessing of Cdks for the treatment of cancer and other diseases [27, 28]. All three D-type cyclins (D1–3) [29] form physical complexes with the retinoblastoma protein (pRb) [22, 30, 31]. A pivotal role for cyclin D1 in fibroblasts [32, 33], myocytes [34], and mammary epithelial cells [35] was demonstrated using antibody immunoneutralization or antisense cyclin D1 expression plasmids. Inhibition of cyclin D1 expression results in cell cycle arrest, whereas its moderate overexpression accelerates G_1 phase progression [33, 36, 37]. Cyclin D1 is rate limiting in growth factor- or estrogen-induced proliferation of mammary epithelial cells [35] and is therefore a critical target for proliferative signals in G_1. The predominant Cdks associated with cyclin D1 are Cdk4 [38] and Cdk6 [39]. Phosphorylation of pRb is critical in modulating G_1-S phase progression and tumor suppressor activity [22]. The sites of pRb phosphorylation mapped in vivo correlate well with those sites recognized by Cdks [40, 41]. The most important cyclin-Cdk complexes that are responsible for phosphorylating pRb during the G_1-S transition are cyclin D1-Cdk4, cyclin D1/Cdk6, and cyclin E/Cdk2 [21, 22]. Cyclin D1-dependent phosphorylation sites in pRb are distinct from the sites phosphorylated by cyclin E-Cdk2 complexes [42]. The biological significance of cyclin D1 in development and tumorigenesis was established through in vivo experiments in the 1990s, using cyclin D1 antisense plasmids [43] and with subsequent gene knockout experiments, demonstrating that cyclin D1 was required for breast [44], skin [45], and gastrointestinal tumorigenesis [8].

3.3   Regulation of Transcription Factor Activity in Cultured Cells

In addition to being the regulatory subunit of Cdk4/6, cyclin D1 regulates the transcriptional activity of a subset of TFs [6]. Cyclin D1 was first shown to repress the activities of several transcription factors, including c-Myc [46], the Myb-like DMP1 transcription factor [47, 48], Neuro D [49], B-Myb [50], Myo D [51, 52], STAT-3 [53], and others [54]. Cyclin D1 has since been shown to either activate or repress more than 40 TFs in gene reporter assays, including the thyroid hormone receptor (TR) [55], the CAAT enhance-binding protein (C/EBPß) [66], the peroxisome proliferator-activated receptor gamma (PPARγ) [57], the estrogen receptor (ERα) [56, 58, 59], and the androgen receptor (AR) [11, 60, 61]. Cyclin D1 was also found associated with the activation domain of STAT-3 upon interleukin-6 stimulation [53]. In this experimental setting, the overexpression of cyclin D1 inhibited the transcriptional activation mediated by STAT-3. This effect was not shared by cyclin E and was independent of the ability of cyclin D1 to bind and activate Cdk4 (it was therefore unaffected by inhibitors of Cdk4).

Mutational analyses of cyclin D1 have been used to identify the protein domains involved in transcriptional regulation, and mutational analyses of target

TFs have further been used to understand the potential mechanisms involved. Based on these experiments, it is clear that distinct domains of cyclin D1 govern its transcriptional repression and transcriptional activation functions. Cyclin D1-mediated repression of MyoD-mediated transcription and muscle differentiation was, at least in part, associated with a pRb-independent mechanism [51, 62]. Likewise, stimulation of ERα-mediated transcription by cyclin D1 involved a direct association between these proteins and also occurred independently of Cdk4, suggesting that cyclin D1 may serve as a direct transcriptional regulator of hormone signaling [58, 59]. Indeed, mutational analyses demonstrated that cyclin D1 interacts directly with the ligand-binding domain of ERα and stimulates ERα transactivation in a ligand- and Cdk-independent fashion [58, 59]. In subsequent studies, cyclin D1 was also shown to bind the ERα coactivator SRC1 [63]. Cyclin D1 interacts, in a ligand-independent fashion, with coactivators of the SRC1 family through a motif that resembles the leucine-rich coactivator-binding motif of nuclear receptors. Cyclin D1 may thus serve as a bridging factor between ERα and SRCs, recruiting SRC-family coactivators to ERα sites in the absence of ligand [63].

A summary of the Cdk-dependent and Cdk-independent functions attributed to cyclin D1 is shown in Figs. 3.2 and 3.3.

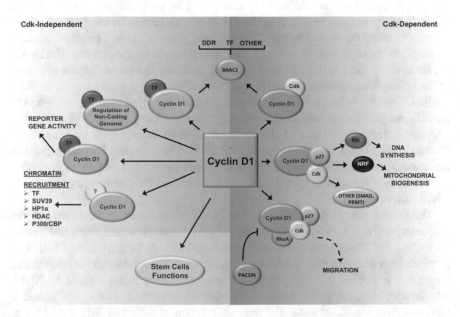

Fig. 3.2 The cellular functions of cyclin D1. Cdk-dependent functions of cyclin D1 (*blue box*) are involved in the regulation of DNA synthesis, mitochondrial biogenesis, cellular migration, and the DNA damage response (*DDR*). On the other hand, Cdk-independent functions (*purple box*) of cyclin D1 regulate gene expression by interacting with transcription factors, as well as through its noncoding genome functions and the recruitment of histone-modifying enzymes. In addition, cyclin D1 has a key role in stem cells function (*orange box*)

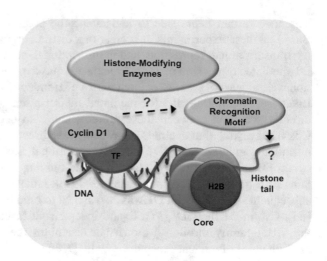

Fig. 3.3 Cyclin D1 can occupy gene promoters and help establish specific histone modifications. Cyclin D1 can gain access to chromatin and gene promoter regions through its ability to interact with various transcriptions factors and recruit histone-modifying enzymes implying the presence of a chromatin recognition motif in cyclin D1

3.4 Transcriptional Regulation of Fat Metabolism

The physiological role of cyclin D1 in restraining fat metabolism is a paradigmatic example of the manner in which cyclin D1 conveys multiple transcriptional effects in order to affect the same function. The binding of cyclin D1 to TFs affects cellular metabolism through several mechanisms, restraining both the activity of TFs and the function of coactivators that govern fat metabolism. The physiological relevance of the interactions between cyclin D1 and these TFs has suggested a model in which cyclin D1 normally functions to inhibit the fat differentiation pathway governed by peroxisome proliferator-activated receptor gamma (PPARγ) and/or C/EBPß [6, 57]. The metabolic consequence of PPARγ repression by endogenous cyclin D1 was revealed in *cyclin D1$^{-/-}$* mice. In these animals, PPARγ signaling was activated with the consequent induction of lipid metabolism and the arising of fatty liver disease [6, 57]. In fact, the liver of *cyclin D1$^{-/-}$* mice resembled the phenotype produced by PPARγ overexpression.

The PPARγ coactivators p300 [12] and PGC1α (PPARγ coactivator-1α) [64] are also restrained by cyclin D1 in ways that are kinase-independent (p300) and kinase-dependent (PGC1α) (described in detail below). The inhibition of PGC1α in particular is associated with inhibition of mitochondrial metabolism. Mitochondrial biogenesis is governed by mtTFA via nuclear respiratory factor 1 (NRF1). NRF1, which induces expression of nucleus-encoded mitochondrial genes, was shown to bind to cyclin D1. In addition, the phosphorylation of NRF1 at Serine 47 [65] by cyclin D1-Cdk complexes represses both the expression and activity of NRF1.

Taken together, these studies are consistent with a model in which cyclin D1 functions to restrain lipid metabolism through both Cdk-dependent and

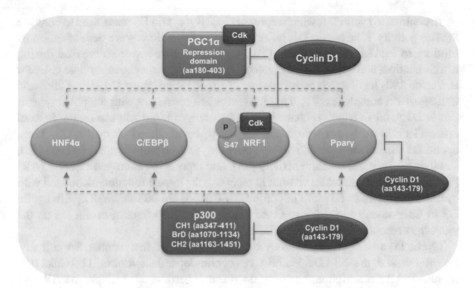

Fig. 3.4 The kinase-dependent and kinase-independent actions of cyclin D1. Cyclin D1 inhibits both PGC1α and NRF1 in a kinase-dependent manner. The repression domain of PGC1α (aa180-403) is phosphorylated by cyclin D1-Cdk4 complexes. Moreover, cyclin D1-Cdk4 complexes directly inhibit NRF1 through phosphorylation in Serine 47 (S47). However, cyclin D1 inhibits the activity of transcription factors and histone acetylases, such as PPARγ and p300, through a kinase-independent mechanism. The domain of cyclin D1 delimited by residues 143–179 plays a crucial role in the repressive action of cyclin D1 on PPARγ and p300. The cyclin D1-mediated repression of p300 also requires specific domains within p300, such as the CH1 (aa347-411), Bromo (BrD, aa1070-1134) and CH2 (aa1163-1451) domains

Cdk-independent mechanisms [7, 8]. Although the findings that the cyclin D1-associated kinase activity can phosphorylate NRF1 [65] and PGC1α could undermine the therapeutic use of Cdk inhibitors in cancer, no such deleterious metabolic consequences have been reported in humans. Figure 3.4 *summarizes both the Cdk-dependent and Cdk-independent functions of cyclin D1 in metabolism.*

3.5   Cyclins D2 and D3

Cyclin D2 was shown to interact with the transcription factor GATA-4 [67]. The polycomb group gene product Mel-18 also binds to the amino terminus of cyclin D2 [68]. On the other hand, cyclin D3 binds to, and negatively regulates the activity of, AML1. Mechanistically, cyclin D3 seems to compete with C/EBPß for AML1 binding, thereby diminishing AML1 target gene affinity and subsequently preventing cell cycle progression [69].

In contrast to cyclin D1, which represses PPARγ, cyclin D3 was found to enhance PPARγ activity. Like cyclin D1, however, cyclin D3 actions were dependent on its binding to PPAREs in chromatin [70]. Cyclin D3 levels were increased during differentiation, and cyclin D3-associated kinase activity can phosphorylate C/EBPα at Serine 193, leading to the formation of growth-inhibitory C/EBPα-Cdk2 and C/EBPα-Brm complexes [71]. These findings are consistent with the phenotype of *cyclin D3⁻/⁻* mice, which is characterized by compromised adipose tissue. Several other TFs and nuclear receptors are known to be regulated by cyclin D3. Thus, cyclin D3 can enhance the activity of activating transcription factor 5 (ATF5) [72], can induce vitamin D receptor activity [73], and repress the androgen receptor (AR) [74]. Regarding its effect on vitamin D receptor, the kinase function associated with cyclin D3 appears to be more important, as Cdk4 is found in complex with cyclin D3 in ChIP assays and Cdk4 or Cdk6 counteracted the effect of cyclin D3 on the vitamin D receptor (VDR).

Cyclin D3 also repressed AR activity in a kinase-dependent manner. Specifically, cyclin D3 and the 58-kDa isoform of cyclin-dependent kinase 11 (Cdk11^{p58}) repressed AR transcriptional activity. In these experimental settings, the AR was phosphorylated by cyclin D3/Cdk11^{p58} on Ser-308 [74] both in vitro and in vivo, leading to the repressed activity of AR's transcriptional activation unit 1 (TAU1). The finding that cyclin D3-mediated AR repression was Cdk-dependent [75] is in contrast with earlier studies suggesting that cyclin D3-mediated repression of AR was Cdk-independent [76]. The differences may be due to different experimental approaches as Cdk independence was based in part on the finding that cyclin D3 repressed AR activity in pRb-negative SAOS2 cells [76].

3.6   D-Type Cyclins and Coactivator Proteins

D-type cyclins are found associated with several intracellular proteins with intrinsic histone acetyltransferase (HAT) activity. For example, cyclin D1 associates with several coactivators with HAT activity, including SRC1 [63], p300/CBP [12], p300/CBP-associated factor (P/CAF) [11, 77, 78], general control non-repressed protein 5 (GCN5) [79], BRCA1 [80], and the basal transcription apparatus-binding protein TAFII 250 [81]. The structure and function of HAT domains are highly conserved [82, 83]. GCN5 and the related P/CAF [84], in particular, are conserved HATs [83] whose activities on nucleosomes facilitate initiation of transcription (reviewed in ref. [85]). It has been shown that cyclin D1 associates with, and modulates the function of coactivators in either a Cdk-dependent or Cdk-independent manner. The N-terminus of PGC-1 coactivators have a strong transcriptional activation domain that interacts with proteins containing histone acetyltransferase (HAT) activity. Through its association with PGC1α, cyclin D1 represses gluconeogenesis and oxidative phosphorylation in part via inhibition of PGC1α activity in a Cdk4-dependent manner [64]. PGC1α is a PPARγ coactivator, and the metabolic phenotype of reduced hepatic fatty acid oxidation in PGC1α knockout, and increased fatty acid oxidation in the cyclin D1$^{-/-}$ mouse, is consistent with prior findings suggesting that cyclin D1 inhibits PPARγ activity [7, 8].

Because a substantial number of TFs are regulated by limiting the abundance of the co-integrator p300 [86], studies were conducted to analyze the genetic interaction between *cyclin D1* and *p300* using gene deletion approaches [12]. Cyclin D1 has been shown to recruit p300 to TF binding sites in ChIP assays [7, 8, 12], and cyclin D1 inhibited TF activity via p300. For example, cyclin D1 inhibited B-Myb activity by interfering with the interaction between B-Myb and p300. p300, a protein with HAT activity, is a co-integrator required for the regulation of multiple transcription factors. It is thought that cyclin D1 can bind p300 and thus repress p300-mediated transactivation [12]. Both cyclin D1 and p300 were demonstrated at the PPARE of the lipoprotein lipase promoter [12]. Importantly, cyclin D1 repressed p300 independently of cyclin D1's ability to interact with Cdk and pRb, and the experiments were consistent with prior findings that p300 augments TF activity and that cyclin D1 restrains activity of these TFs. In addition, consistent with findings that cyclin D1 inhibited HAT activity of P/CAF [77], cyclin D1 also inhibited HAT activity of p300 in vitro [12]. The functional relationship between p300 and cyclin D1 was supported by *p300* and *cyclin D1* genetic deletion analyses. Microarray analysis also identified a set of genes that were repressed by endogenous cyclin D1 and maintained by p300. These genes were involved in cellular differentiation and induction of cell cycle arrest [12]. Mutational analysis of the interaction between cyclin D1 and p300 revealed that the bromodomain and cysteine- and histidine-rich domains of p300 were required for the repression mediated by cyclin D1 [12]. The, deletion of amino acids 143–178 of cyclin D1 abolished repression of p300. The same region of cyclin D1 is referred to as the "repression domain" as it was originally shown to repress PPARγ [57] and AR activity, and is predicted to form a helix-loop-helix structure [57]. Further mechanistic studies demonstrated that cyclin D1-mediated inhibition of HAT activity primarily affected p300 auto-acetylation and p300-mediated acetylation of H4 and H2A/B [12].

P/CAF associates with cyclin D1 and AR through similar domains [77]. Therefore, cyclin D1 can disrupt the binding of AR to P/CAF in vitro. These findings were consistent with a model in which cyclin D1 restrained AR activity via reduction of the HAT activity of P/CAF. Several lines of evidence have thus supported a model in which cyclin D1 governs TF activity via modulation of HAT function. Studies carried out in Mark Ewen's laboratory demonstrated that P/CAF potentiates cyclin D1-mediated stimulation of ERα activity in a dose-dependent manner, a finding that was largely dependent upon the acetyltransferase activity of P/CAF [78].

The product of the *BRCA1* breast cancer susceptibility gene functions as both a tumor suppressor protein and transcriptional regulator [87–89]. Among a variety of transcriptional functions, BRCA1 interacts with components of the histone deacetylase complex [90] and was also shown to repress the ligand-dependent transcriptional activity of ERα [91, 92]. Binding of cyclin D1 to BRCA1 antagonizes BRCA1-

mediated ERα repression, with the consequent induction of ERα signaling [80]. This effect of cyclin D1 on BRCA1 at the estrogen response element (ERE) was independent of cyclin D1-associated kinase function and seemed to depend on BRCA1's recruitment to chromatin [80, 93].

As noted before, the activity of PPARγ is enhanced by PGC1α. PGC1α, in turn, controls metabolism in a variety of tissues [94], activating mitochondrial metabolism partially through induction of NRF1 [95]. Interestingly, PGC1α is repressed by cyclin D1 through an intriguing mechanism [79]. As noted above, the HAT activity of GCN5 was enhanced by cyclin D1-Cdk complexes through phosphorylation of Threonine 272 and Serine 372, which in turn acetylated and thereby repressed PGC1α [79]. These studies were consistent with earlier studies on cyclin D1 function in the liver, which revealed repression of PPARγ signaling together with hepatic steatosis in *cyclin D1^{-/-}* mice. This phenotype was previously attributed to cyclin D1-mediated inhibition of PPARγ function, which was shown to occur via direct protein-protein interaction and was largely mediated by the repression domain of cyclin D1 (amino acids 143–179). Subsequent studies have revealed several additional ways in which cyclin D1 restrains PPARγ signaling. PGC1α binds and is inactivated by Cdk4-mediated phosphorylation at Threonine 298 and Serine 312. Thus PGC1α is a novel cyclin D1-Cdk4 substrate [64]. Studies by Puigserver et al. showed that cyclin D1-Cdk4 complexes phosphorylate and thereby activate GCN5, which in turn acetylates and thus inhibits PGC1α activity. In addition, studies by Bhalla [64] and Puigserver [79] showed that cyclin D1 inhibited gluconeogenic enzymes. It is likely that cyclin D1-mediated inhibition of mitochondrial metabolism [65, 96] may also involve the inhibition of PGC1α.

Transcription factors regulated by PGC1α (MEF2a, HNF4α, NRF1, PPARγ) may be repressed by cyclin D1 through PGC1α inactivation. In this regard, prior studies from Jeffrey Albrecht's laboratory had shown that cyclin D1 inhibits hepatic lipogenesis in part via repression of the carbohydrate response element-binding protein (ChREBP) and HNF4α [97]. Collectively, these studies are consistent with a model in which cyclin D1 restrains PPARγ signaling at multiple levels, inhibiting the coactivators p300 and PGC1α as well as ChREBP and HNF4α.

3.7   Cyclin D1 Changes the Association of TFs to Chromatin

Several studies have shown that cyclin D1 may either enhance or restrain the recruitment of TFs to target DNA sequences. In studies carried out in cyclin D1-deficient 3T3 cells, PGC1α recruitment to its cognate binding site (PPARE) at the lipoprotein lipase (LPL) promoter was dependent on the abundance of endogenous cyclin D1 [8]. In addition, the levels of endogenous cyclin D1 influenced the recruitment of ERα to EREs [80], increasing the abundance of ERα at these elements [80]. Cyclin D1 also reduced the recruitment of the corepressor BRCA1 to pS2's promoter sites (EREs) [80] and governed the recruitment of p300 to LPL promoter sites (PPAREs) [12].

In other studies, cyclin D1 reduced KLK2/PSA (kallikrein-related peptidase 2/prostate-specific antigen) expression, and AR binding to the KLK3/PSA and transmembrane protease, serine 2 (TMPRSS2) ARE promoter sites was reduced by cyclin D1 both at the basal state and in the presence of DHT [98]. In vivo, however, cyclin D1 enhanced DHT-induced abundance of the murine homologue of KLK3/PSA in the prostate [99] suggesting differences between the human and murine systems and/or technical differences between different laboratories. Collectively, these studies have demonstrated that endogenous cyclin D1 may determine the recruitment of TFs to their cognate binding sites in the context of chromatin.

3.8   Cyclin D Regulation of the Basal Transcription Apparatus

$TAF_{II}250$, the largest subunit of the TFIID complex, was first identified as the cell cycle control gene, *CCG1* [100–102]. The cyclin H-Cdk7 complex, also known as cyclin activating kinase (CAK), is associated with TFIIH and phosphorylates the carboxyl terminal domain (CTD) of the largest subunit of RNA polymerase II in order to initiate transcriptional elongation [103]. Robbins and colleagues showed that Sp1-mediated transcription is stimulated by pRb [104] and is repressed by cyclin D1 [105]. In later studies, cyclin D1 was found in association with the TBP-associated factor $TAF_{II}250$, where it contributes to the regulation of Sp1-mediated transcription [81]. The amino terminus of cyclin D1 (amino acids 1–100) was sufficient for its association with $TAF_{II}250$.

3.9   Changes in Local Chromatin Associated with D-Type Cyclins

Cyclin D1 modifies histone acetylation of local chromatin around the binding sites of specific transcription factors. Transcriptionally active chromatin correlates with deacetylation of histone H3K9 [9, 10]. The deacetylation of histones may be mediated through recruitment of histone deacetylases. The histone deacetylase HDAC1 was recruitment by cyclin D1 to the promoter of the LPL gene [7] and Neuro-beta 2 gene [108]. Cyclin D1 has been found associated with a transcriptional repression complex that comprises HDAC1, HDAC3, SUV39, and HP1α [7, 12]. Thus, cyclin D1 may change the posttranslational status of histone, including H3K9 (reducing acetylation, increasing di-methylation) [8] and H3K4 (reducing di-methylation). Cyclin D1 can either induce or repress gene expression. Specific types of histone acetylation occur at local cis elements corresponding to either repression or activation [106, 107]. The recruitment of cyclin D1 to target genes assessed to date shows a correspondence between the cyclin D1-mediated change in gene expression and the anticipated changes in histone acetylation. Cyclin D1 abundance alters local histone acetylation at promoter regulatory regions, thereby determining histone H3 acetylation at K9, di-methylation of H3K4, and di-methylation of H3K9 [7, 12].

The relative abundance of cyclin D1 at transcriptional regulatory elements of promoters is also regulated. Cyclin D1 is rate limiting for the recruitment of PPARγ to the PPARE binding sites of the LPL promoter (in vivo) [8]. During differentiation of adipocytes, cyclin D1 occupancy becomes reduced at the regulatory regions of those genes that are induced. As the LPL gene is induced during differentiation, cyclin D1 levels at the PPARE decrease, concomitant with a decline in the repression complex at the PPARE. In this manner, the relative abundance of cyclin D1 within the cell may coordinate expression of genes during adipogenesis.

Transcriptionally active chromatin correlates with deacetylation of histone H3K9 [9, 109]. Our studies [8] demonstrated that cyclin D1 was recruited to the PPARγ response element of the LPL promoter, and this event coincided with reduced acetylation of histone H3K9. Subsequent studies demonstrated the presence of cyclin D1 in chromatin complexes at the AP-1 site of the cyclin D1 promoter [15]. ChIP studies using *cyclin D1*$^{-/-}$ cells that were rescued with physiological levels of human cyclin D1 demonstrated the recruitment of cyclin D1 to the cyclin D1 promoter. In addition, cyclin D1 was identified as part of a complex at AP-1 and CRE sites of the cyclin D1 promoter [8].

3.10   Cyclin D1 Interacts with Histone Methylases

The recruitment of cyclin D1 to chromatin was also accompanied by the recruitment of the histone-lysine N-methyltransferase Su(Var)39H1 and the heterochromatin protein HP1α to chromatin. Furthermore, cyclin D1 was shown to promote the coordinated recruitment of both HDAC1 and HDAC3, together with HP1α and SUV39, and these events correlated with a reduction in acetylation of histone H3K9. Su(Var)39H1 is a H3K9 methylase which plays a vital role in heterochromatin organization, chromosome segregation, and mitotic progression [110]. Thus, cyclin D1-mediated recruitment of SUV39H raises the possibility of a link between cyclin D1 and chromosomal function.

In addition to recruiting histone methyltransferases, cyclin D1 actively regulates methyltransferase activity. Arginine methylation is catalyzed by a family of enzymes known as protein arginine methyltransferases (PRMTs) [111]. Among other protein subunits, MEP50 is a WD40 repeat-containing protein that contributes to PRMT5 activity. It has been shown that cyclin D1-Cdk4 complexes phosphorylate MEP50, leading to an enhancement of PRMT5 activity. PRMT5 and its cofactor, MEP50, had been previously identified as novel substrates of the oncogenic cyclin D1(T286A)-Cdk4 complexes [112]. PRMT5/MEP50 activation by cyclin D1(T286A)-Cdk4 results in high levels of p53 methylation and a reduction in p53-dependent apoptosis. PRMT1 is important for lymphomagenesis and elegant studies carried out at Dr. Diehl's laboratory showed that overexpression of PRMT5 cooperated with cyclin D1 in driving mouse lymphomagenesis, a phenotype that was associated with p53 methylation and reduced apoptosis. Importantly, analysis of human tumor specimens revealed a strong correlation between cyclin D1 overexpression and p53 methylation, supporting the pathological relevance of this pathway [113]. These findings are consistent with prior studies in which E2F1-mediated apoptosis was attenuated by PRMT5. PRMT5-dependent methylation thus favors proliferation by antagonizing methylation of E2F1, ultimately reducing E2F1-mediated apoptosis [114].

3.11   Genome-Wide Binding Studies of Cyclin D1 in Chromatin

In view of the abovementioned findings that cyclin D1 can occupy the regulatory regions of promoters and modify chromatin structure through its association with other regulatory proteins, such as p300 [8, 12], genome-wide analyses of cyclin D1 occupancy were conducted. In order to determine sites of cyclin D1 binding, *cyclin D1$^{-/-}$* MEFs were transduced with an expression vector encoding a FLAG epitope-tagged cyclin D1 [17]. Approximately 2840 genes were identified whose promoter regions (within 10 kb of the start sites) bound cyclin D1. Peak values of active regions within promoters were comparable to those 10 kb and beyond, consistent with a model in which cyclin D1 interacts with both proximal and distal elements. ChIP-Seq analysis confirmed the identity of TFs enriched in the cyclin D1 peak interval. These TFs included CTCF, ELK4, SP1, E2F1, ESR2, HOXA5, KLF4, SRF, PPARΓ, BRCA1, HIF1α and NF-κB, C2H2-type zinc-finger protein family, ESR1, and estrogen receptor R1 (also known as ERα). A substantial overlap of binding sites with those identified by Bienvenu et al. was also evident [115]. Because cyclin D1/Cdk-mediated phosphorylation of pRB [116] and related proteins leads to the release of E2F proteins [117–120] and E2F sites abound in the genome, it was important to compare the relative occupancy of E2F sites. The enrichment for E2F1 was significantly less than the enrichment for cyclin D1-associated transcription factors, and the *P* value for E2F1 association in cyclin D1 ChIP was two orders of magnitude lower than the P value for ESR1 (ERα) and three orders of magnitude lower than the P value for CTCF. These findings suggest that cyclin D1-dependent regulation of E2F1 signaling, which is also Cdk-dependent [121], provides only a modest contribution to this particular transcriptional signaling activity.

   To gain a better understanding of the activities regulated by cyclin D1 once recruited into the chromatin, analyses of functional pathways were conducted. Functional pathways that could be inferred from the analysis of promoter regions to which cyclin D1 was bound included those involved in mitochondrial function, and DNA processes such as chromosomal organization and genomic/chromosomal stability [17]. In particular, cyclin D1 seems to repress spindle checkpoint control genes and mitochondrial genes, consistent with studies in fibroblasts and mammary and liver epithelial cells [65, 96].

   ChIP analyses confirmed cyclin D1 occupancy at the regulatory regions of several genes involved in chromosomal stability [17]. Chromosomal instability (CIN), a hallmark of tumor cells, is characterized by chromosomal abnormalities as well as by an altered gene expression signature. CIN is also characterized by an elevated rate of gain or loss of whole chromosomes (aneuploidy) and/or structural chromosomal aberrations [122–124]. One of the most striking differences between cancer cells and their normal counterparts is indeed aneuploidy. While the molecular mechanisms responsible for CIN in tumors remain poorly understood [125, 126], cell cycle-associated proteins, including cyclin E, have been implicated [127]. The relative enrichment of CIN-related genes (including, *AURKB*, *TOP2A*, *CENPP*, *MLF1IP*, *ZW10*, and *CKAP2* [125]) in ChIP and other expression analyses has been used to quantitate CIN [128]. Interrogation of gene expression in 2254 breast tumors identified the expression of cyclin D1 as a strong indicator of CIN in luminal B breast cancer [17]. Importantly, the introduction of cyclin D1 into *cyclin D1$^{-/-}$* MEFs induced a gene expression profile that was reminiscent of the expression profile observed in CIN and that observed secondary to the occupancy of CIN-associated regulatory regions by cyclin D1. Accordingly, mammary gland-targeted cyclin D1 expression led to the formation of tumors characterized by CIN, and short-term transgenic expression of cyclin D1 also led to CIN in vivo. The induction of expression of CIN genes upon cyclin D1 expression was independent of cyclin D1's kinase binding domain [129, 130].

3.12   Functional Interactions Between Cyclin D1 and Transcription Factors In Vivo

Several functional interactions between cyclin D1 and TFs have been assessed for their biological significance in vivo. So far, the interactions between cyclin D1 and PPARγ, DMP1, ERα, C/EBPβ, and AR have been examined in vivo. The interaction with PPARγ was already described under "transcriptional regulation of fat metabolism."

One of the first TFs shown to bind cyclin D1 was cyclin D-interacting myb-like protein 1 (designated DMP1) [47, 48]. DMP1 could bind cyclin D1 and was also phosphorylated by cyclin D-dependent kinases. DMP1 binds the DNA consensus sequence CCCG(G/T)ATGT in order to activate transcription. It was shown that DMP1 activity was antagonized by cyclin D1, an effect that was independent of cyclin D1-associated kinase activity [48]. In subsequent studies, DMP1 was shown to be an haploinsufficient tumor suppressor protein that, upon reduced expression, accelerated Myc-induced lymphomagenesis with a concomitant reduction in the mutational rates of p53 within the tumors [131]. Regarding the functional interaction between DMP1 and cyclin D1 in vivo, a cooperation between cyclin D1 loss and DMP expression was observed in breast cancer [132]. DMP1 activates both the *Arf* and *Ink4a* promoters and, consequently, induces apoptosis or G_2/M cell cycle delay in normal cells [132]. Cyclin D1-induced *Ink4a/Arf* expression was indeed dependent on DMP1, since induction of *Ink4a/Arf* expression was not detected in *DMP1*-deficient or DMP1-depleted cells [132]. *Arf/Ink4a* expression was increased in premalignant mammary lesions derived from *DMP1⁺/⁺*, *MMTV-cyclin D1*, and *DMP1⁺/⁺*, and *MMTV-D1 (T286A)* mice but was significantly downregulated in those lesions derived from *DMP1*-deficient mice. Selective *DMP1* deletion was found in 21% of the *MMTV-D1-* and *MMTV-cyclin D1 (T286A)*-driven mammary carcinomas, and *DMP1* heterozygous status significantly accelerated mouse mammary tumorigenesis. Recently, the *DMP1* locus was shown to generate two splice variants, a tumor-suppressive DMP1α (p53-dependent) and an oncogenic DMP1β (p53-independent). The DMP1β/DMP1α ratio seems to increase with the neoplastic transformation of breast epithelial cells [133]. This process is associated with high DMP1β protein expression and a shorter survival of breast cancer patients. Like *DMP1, ARF* is also frequently inactivated by aberrant splicing in human cancers [134]. The functional significance of these alternatively spliced forms and their role in transcription remains to be explored.

Early studies had shown that cyclin D1 can bind to C/EBPβ and augment C/EBPβ-dependent reporter activity, contributing to the regulation of a common gene signature in human breast cancers [66]. Cyclin D1 binds the C/EBPβ isoform, LAP1, and this interaction leads to an activation of the transcriptional function of LAP1, after relieving its auto-inhibited state [135]. In addition, cyclin D1 and C/EBPβ co-localized to the CEBP site of the *HSC70* promoter in differentiated mammary epithelial cells. Re-expression of *LAP1* restored the ability of *C/EBPβ*-deficient mammary epithelial cells to differentiate and did so in a manner that was dependent on cyclin D1.

Genetic studies have confirmed the biological significance of cyclin D1 in both ERα signaling in the mammary gland [136] and AR signaling in the prostate gland [99]. The functional interaction between ERα signaling and the transcriptional function of cyclin D1 has also been examined in vivo. Cyclin D1 is induced by ERα, but not ERβ, via transcriptional mechanisms [137]. Early studies showed that binding of cyclin D1 to ERα led to an increase in ERE reporter gene activity in a Cdk-independent manner [58, 59]. Cyclin D1 activates estrogen receptor-mediated transcription in the absence of estrogen and enhances transcription in its presence [58], partially through the recruitment of the ERα coactivator SRC1 [63]. In addition, the BRCA1-mediated transcriptional repression of ERα [91, 92] was antagonized by cyclin D1 through physical association in cultured cells and through recruitment of BRCA1 to EREs [80]. Subsequent studies were set out to determine the biological significance of cyclin D1 in global ERα signaling. Expression profiling of 17β-estradiol-stimulated MCF7 cells that were subjected to small interfering RNA (siRNA)-mediated knockdown of cyclin D1 showed that cyclin D1 was required for estrogen-mediated gene expression in vitro [136]. More recent studies have examined the functional significance of cyclin D1 to ERα signaling in vivo using the mammary gland of *cyclin D1*-deficient mice as a model. Genome-wide expression profiling of 17β-estradiol-treated, castrated, virgin mice that were also deficient in cyclin D1 demonstrated that cyclin D1 modulates estrogen-dependent gene expression of more than 80% of estrogen-responsive genes in vivo [136]. The cyclin D1-dependent estrogen signaling pathways identified in vivo were highly enriched for growth factor receptors (EGFR, ERBB3, and EPHB3) and their ligands (amphiregulin, encoded by *AREG*), as well as matrix metalloproteinases [136]. Of note, the ERα can be found in different compartments of the cell, including the membrane, mitochondria, and the nucleus [138]. The non-genomic, extranuclear functions of estradiol can be assessed using a nuclear-excluded estradiol-containing dendrimers. Curiously, part of cyclin D1-dependent ERα signaling was induced using a nuclear-excluded estradiol-containing dendrimer suggesting an additional layer of complexity in cyclin D1-dependent ERα signaling pathway [139].

Unlike most TFs (listed above), for which consistent findings have been reported by multiple laboratories, there are still conflicting reports on the role of cyclin D1 in AR-dependent gene expression and function. Early studies showed that cyclin D1 overexpression in prostate cancer cells (LNCaP) promoted proliferation and tumorigenicity [140]. Consistent with these findings, genetic deletion of cyclin D1 in the mouse reduced DHT-dependent cellular proliferation of prostate cells in vivo [99]. Furthermore, cyclin D1 depletion reduced androgen-induced proliferation of LNCaP cells in vitro [141], as well as of PCa cells in vivo [142]. Similarly, the genetic deletion of cyclin D1 reduced prostate epithelial cell proliferation in vivo [141]. Overall, endogenous cyclin D1 seemed to affect the ability of the AR to modulate the expression (both induction and repression) of >90% of AR target genes [99]. Importantly, microarray analyses highlighted the importance of the gene network centered around cyclin D1 to promote prostate stem cell expansion via a Wnt/β-catenin signaling pathway [99]. These findings are consistent with prior studies from the DePinho's laboratory showing that cyclin D1 expression correlated with poor prognosis of patients with prostate cancer [143]. Interestingly, a cyclin D1-centered gene expression signature was able to sort patients with poor therapeutic outcome, with a power that was better than the use of cyclin D1 protein levels [141]. Furthermore, numerous studies have shown that factors that induce cyclin D1 expression in prostate cancer cells also induce cell proliferation [12, 143–146]. In contrast with these in vivo and in vitro studies, Dr. K. Knudsen et al. showed that cyclin D1 inhibited DNA synthesis in LNCaP cells [147], a finding that may be related to differences in experimental approaches.

Cyclin D1 mRNA is alternatively spliced to yield two different transcripts, which are translated into two functionally different proteins [148, 149]. It has been shown that an A870G polymorphism in Exon 4 of CCND1 is crucially involved in this alternative splicing [148]. So far, however, there have been contrasting reports as to what the effects of each isoform (referred to as cyclin D1a and cyclin D1b) on AR function and prostate cancer risk, might be. One early study suggested that the A870 polymorphism, known to facilitate production of cyclin D1b, correlated with poor outcome. Importantly, cyclin D1b failed to repress the AR [147]. Subsequently, Wang et al. reported that the A allele of the A870G polymorphism was associated with susceptibility to prostate cancer [150], but no such association was found by Chen et al. [151]. A further meta-analysis of 3820 cases and 3825 controls did not find any differences in prostate cancer outcome [152]. Similarly, no differences in the ability to repress AR were found between cyclin D1a and cyclin D1b, at least based on gene reporter assays [139]. Thus, although cyclin D1 was able to repress

AR gene reporter activity [11, 61], gene expression analyses using cyclin D1 knockdown approaches, as well as in vivo studies, have shown that cyclin D1 can modulate both AR repression and AR activation [99, 141]. Given the dramatic change in gene expression [99] and the reduction in proliferation observed in cyclin D1-deficient prostate cells in the presence of androgens, it will be important to further examine the functional interaction between cyclin D1 and AR.

3.13   Cyclin D Regulation of the Noncoding Genome

Thus far, relatively little is known about the mechanisms through which the cell cycle proteins, including cyclins, regulate the noncoding genome. Micro-RNAs (miRNAs) are 21- to 22-nucleotide-long molecules that modulate a variety of cellular phenotypes by affecting the translational efficiency or the stability of targeted mRNAs. Today, there is compelling evidence for the importance of the noncoding genome and the miRNA biogenesis apparatus in tumorigenesis. A comparative analysis of miRNA expression in cyclin D1-induced mammary tumors and mammary tissues derived from cyclin D1 antisense or knockout mice identified the miR-17/20 cluster as a cyclin D1-induced regulator of mammary tumor growth. miR-17/20 can repress the expression of cyclin D1 by targeting the 3' untranslated region of its mRNA [153]. Using ChIP assays, cyclin D1 was found associated to the miR-17/20 regulatory region, between nucleotides −1050 and −1200. To our knowledge, these were the first studies that demonstrated cyclin-dependent regulation of a noncoding RNA through binding to the regulatory region of a miRNA cluster [154]. miR-17/20, in turn, regulates the secretion of cytokines and plasminogen activator via the expression of α-enolase and cytokeratin 8. The inhibition of the plasminogen activator by miR-17/20 required cyclin D1, indicating that complex regulatory loops between the noncoding and the coding genome are involved in the regulation of migration of breast cancer cells [154]. Following the discovery of this regulatory circuit, several studies have found that cyclin E may also be regulated by several distinct miRNAs, including miR-223, miR-161, and miR-195. However, so far there is no evidence that cyclin E can bind to regulatory regions of the noncoding genome in order to coordinate miRNA expression.

Cyclin D1 was also shown to modulate miRNA biogenesis through the transcriptional induction of *DICER* [155]. Dicer is an enzyme that cleaves double-stranded RNA or stem-loop-stem RNAs into 20–25-nucleotide-long small RNAs, a process that is crucial for miRNA maturation. Interestingly, *cyclin D1$^{-/-}$* cells are defective in pre-miRNA processing, which is restored by the reintroduction of cyclin D1a. Cyclin D1 induces the expression of *DICER* in vitro and in vivo [155], and this process seems to be independent of Cdk activation.

Taken together, cyclin D1 can regulate both the expression of individual miRNAs (via binding to their regulatory regions) and can also modulate the processing of other miRNAs via the transcriptional induction of *DICER* [155].

3.14   Transcription and DNA Repair

Numerous studies have now demonstrated the functional co-opting of components of the DNA repair and gene transcription processes (reviewed in [156]). Data extracted from the nuclear receptor signaling atlas have unveiled a substantial overlap between these two processes, and reciprocal co-immunoprecipitation experiments suggested more than 2500 unique pairwise associations [157]. Proteins with clear roles in both transcription and DNA repair include topoisomerase II [158], PARP (Poly(ADP-ribose)), TIP60, BRCA1, BRCA2, and the BRCA2-binding protein P/CAF. In particular, PARP-1 binds to ~90% of RNA polymerase II promoters and, among other functions, promotes ERα activity [158]. Interestingly, cyclin D1 has been shown to interact with, and regulate the transcriptional activity of, several of these dual-function proteins.

BRCA1 is part of the RNA polymerase II holoenzyme. In addition, BRCA1 is dynamically regulated by DNA damage signals [159, 160] and is required for transcription-coupled repair following oxidative DNA damage [161]. Moreover, BRCA1 is a transcriptional repressor of ligand-dependent ERα [91, 92], a role that is partially mediated by acetylation and ubiquitylation of the ERα [162]. As mentioned before, cyclin D1 antagonizes BRCA1-mediated repression of ERα-dependent gene expression [80]. Cyclin D1-mediated repression of BRCA1 was independent of its Cdk, pRb, or SRC1 functions in breast and prostate cancer cells. As cyclin D1 abundance is regulated by oncogenic and mitogenic signals, the antagonistic role of cyclin D1 on BRCA1-mediated ERα may contribute to the selective induction of BRCA1-regulated target genes [80]. The p300/CBP-associated factor (P/CAF), which binds BRCA2 in a phosphorylation dependent manner [163], also binds cyclin D1 [11, 78].

The cell type-specific sensitivity to radiation can also be modulated by the relative abundance of cyclin D1. Gamma radiation-induced apoptosis was enhanced in *cyclin D1*$^{-/-}$ MEFs [164], and cyclin D1 overexpression inhibited UV-induced apoptosis in a p300-dependent manner. The cyclin D1a isoform induces the expression of genes involved in DNA replication and/or the DNA damage checkpoint in fibroblasts and mammary epithelial cells [7, 96]. Cyclin D1, for example, induces the expression of mini-chromosome maintenance-deficient 2 (MCM2), MCM3, and MCM4 [12, 96], whereas cyclin E/Cdk complexes regulate the loading of MCM onto chromatin [165] through both kinase-dependent and kinase-independent mechanisms [166]. In contrast, breast cancer cell lines show enhanced apoptosis in response to gamma radiation when cyclin D1 is overexpressed [167, 168]. Cyclin D1 seems to be important for the G_1 cell cycle arrest induced by gamma radiation because interference with the degradation of cyclin D1 prevents both G_1 arrest and G_2-M arrest in cells subjected to gamma irradiation [169]. Intriguingly, lymphoid compartment-targeted expression of a *cyclin D1 (D1T286A),* which is confined to the nucleus during S phase, was accompanied by an increase of aneuploidy in lymphoid tumors in which a DNA damage response (DDR) had been triggered [170].

We [80, 139, 141, 142, 171] and others [16] have shown that cyclin D1 regulates DNA damage repair [171] and binds DNA repair proteins, which include RAD51 [171] and BRCA1 [80]. Elegant studies carried out in Peter Sicinski's lab have confirmed the ability of cyclin D1 to bind Rad51 and have revealed additional interactions between cyclin D1 and other DNA repair proteins, most importantly BRCA2 [16]. Following the identification of RAD51 as an interacting partner of cyclin D1, and through the use of an homologous recombination repair reporter system, endogenous cyclin D1 was shown to increase the homologous recombination rate. BRCA2 was also identified as a cyclin D1-interacting protein, and BRCA2 knockdown reduced cyclin D1 recruitment to sites of DNA damage. Cyclin D1 depletion, however, did not affect BRCA2 recruitment, although it reduced the recruitment of RAD51 to DNA damage sites [16]. These observations are consistent with previous findings indicating that several transcriptional proteins involved in DNA repair [156] do interact with cyclin D1.

Because DNA repair factors must be tethered to chromatin as part of the DNA damage response (DDR), and cyclin D1 is also recruited to the local chromatin – Li et al. [171] examined the role of cyclin D1 in regulating the DNA damage signaling response. By using comet assays, in which the tail of the comet is used as a surrogate for damaged DNA at a neutral pH, the authors were able to demonstrate that cells carrying wild-type cyclin D1 had a fourfold increase in comet tail formation compared with cyclin D1-deficient cells. Thus, the levels of phosphorylation of H_2AX on serine 139 (γH_2A), a sensitive marker of double-strand breaks, increased in *cyclin D1$^{+/+}$* cells treated with doxorubicin compared to *cyclin D1$^{-/-}$* cells subjected to the same treatment. As expected, siRNA-mediated reduction of endogenous cyclin D1 led to reduced 5-fluorouracil-induced H_2AX phosphorylation. The enhancing effect of cyclin D1 on the DDR occurred rapidly (in 15 min), thus preceding its effect on DNA synthesis (>6 h). Interestingly, though cyclin D1a enhanced the DDR induced by doxorubicin, it did not enhance the entry into S phase in the absence of serum, suggesting that the induction of the DDR by cyclin D1a can be uncoupled from the induction of DNA synthesis.

Cyclin D1 was also shown to bind p21^{Cip1} in addition to RAD51, and cyclin D1-mediated induction of the DDR seems to require p21^{Cip1}. Of note, cyclin D1a expression recapitulated the recruitment of H_2AX in a manner similar to recruitment to DNA damage proteins ATM, NBS1, and MDC1. Mutational analyses demonstrated a requirement for the cyclin D1 carboxyl terminus in its recruitment to the H2AX foci. Cyclin D1a was also shown to recruit RAD51 in the context of local chromatin in response to the DNA damage [171].

3.15 D-Type Cyclins in Distinct Locations: Potential Roles in Transcription

Gene transcription of a significant part of the genome, involving both coding and noncoding regions, must be tightly coordinated during the cell cycle through mechanisms that are, thus far, poorly understood. As noted above, certain E2F transcription factors are coordinately expressed. Accordingly, they also regulate gene expression in a cell cycle-dependent manner. More recently, the cell cycle genes homology region (CHR) has been identified as a DNA element with an important role in transcriptional regulation of late cell cycle genes [172].

On a more general level, we postulate the existence of mechanisms by which topologically distinct pools of D-type cyclins communicate transcriptional information at different times during the cell cycle. For example, D-type cyclins can associate with proteins present at the plasma membrane. Early studies showed that cyclin D1 modulates migration of macrophages [173], fibroblasts [174], and mammary epithelial cells [175]. In fibroblasts, cyclin D1 was shown to modulate the small GTPase RhoA, suggesting a role for cyclin D1 in membrane-associated processes. Mass spectrometry also identified cyclin D1 association with PACSIN II (protein kinase C and casein kinase substrate in neurons 2), and the functional interaction between the two proteins was demonstrated to play a key role in migration [176]. PACSIN family members (also called syndapins) function as cytoplasmic adaptor proteins at sites of focal adhesions, and interestingly, cyclin D1 can also be allocated to focal contacts. While signaling pathways originating in focal contacts have been well described [177, 178], the potential roles of D-type cyclins in these pathways remain unknown. As PACSIN proteins can interact with a great variety of molecules associated to cell membranes (including synaptojanin, dynamin N-WASP) [179], the possibility exists that cyclin D might also play a role as a modulator of membrane trafficking. Of note, the E2-dependent DNA damage signaling that is dependent on cyclin D1 involves an extranuclear (non-genomic) function [139], which appears to involve a membrane-associated form of the ERα.

Importantly, Powers et al. have demonstrated the existence of four distinct nuclear D-type cyclin compartments in pro-B cells, including a CDK4-associated cyclin D3 fraction and a PI3K-regulated fraction that is not required for proliferation. A third fraction of cyclin D3 was associated with the nuclear matrix and seems to be involved in the repression of more than 200 genes, including a subset of variable (V) genes [180]. Consistent with the existence of different subnuclear compartments and functions, distinct domains of cyclin D3 mediated proliferation and gene repression.

The nuclear lamina (NL) interacts with genomic regions known as lamina-associated domains (LADs). Recent work suggests that these contacts are linked to H3K9 di-methylation, which is in turn dependent on the H3K9 methyltransferase G9a [181]. In this manner, G9a contributes to the dynamic architectural changes of chromosomes and gene regulation. As LADs are found on all chromosomes and cover approximately 40% of a mammalian genome, it has been suggested that the interactions of nuclear lamina with LADs may impose specific constraints on the positioning of chromosomes. The mechanisms that drive nuclear lamina interactions with LADs involve long (GA)n repeats [182], and methylation of histone H3 lysine 9 was found to be important for the lamina-mediated anchoring of certain genes in C. *elegans* [183]. Of note, the wide range of interactions between chromatin components and nuclear lamina proteins has been proposed to play an important role in a variety of diseases [184]. Whereas the cyclin D1-binding proteins, HP1α and the Suv39 methylase, enable the coordination of heterochromatin spread and therefore are crucial to gene expression [110], it remains to be determined whether or not D-type cyclins can themselves associate with nuclear lamina and contribute to broad transcriptional regulatory changes.

Acknowledgments This work was supported in part by 1R01CA137494 R01 CA 132115-05A1 (R.G.P.). The Sidney Kimmel Cancer Center was supported by the NIH Cancer Center Core Grant P30CA56036 (R.G.P). This project is funded in part from the Breast Cancer Research Foundation (R.G.P), Dr. Ralph and Marian C. Falk Medical Research Trust (R.G.P), and a grant from Pennsylvania Department of Health (R.G.P.). The department specifically disclaims responsibility for analyses, interpretations, or conclusions.

We sincerely apologize to any authors whose contributions to this field have not been fully described due to the limitations of size for this chapter.

References

1. Pestell RG, Albanese C, Reutens AT, Segall JE, Lee RJ, Arnold A. The cyclins and cyclin-dependent kinase inhibitors in hormonal regulation of proliferation and differentiation. Endocrine Rev. 1999;20(4):501–34.
2. Bregman DB, Pestell RG, Kidd VJ. Cell cycle regulation and RNA polymerase II. Front Biosci. 2000;5:D244–57.
3. Coqueret O. Linking cyclins to transcriptional control. Gene. 2002;299(1–2):35–55.
4. Wang C, Fu M, Mani S, Wadler S, Senderowicz AM, Pestell RG. Histone acetylation and the cell-cycle in cancer. Front Biosci. 2001;6:D610–29.
5. Fu M, Wang C, Wang J, Zafonte BT, Lisanti MP, Pestell RG. Acetylation in hormone signaling and the cell cycle. Cytokine Growth Factor Rev. 2002;13(3):259–76.
6. Fu M, Wang C, Li Z, Sakamaki T, Pestell RG. Minireview: Cyclin D1: normal and abnormal functions. Endocrinology. 2004;145(12):5439–47.
7. Fu M, Rao M, Bouras T, Wang C, Wu K, Zhang X, Li Z, Yao TP, Pestell RG. Cyclin D1 inhibits peroxisome proliferator-activated receptor gamma-mediated adipogenesis through histone deacetylase recruitment. J Biol Chem. 2005;280(17):16934–41.
8. Hulit J, Wang C, Li Z, Albanese C, Rao M, Di Vizio D, Shah S, Byers SW, Mahmood R, Augenlicht LH, Russell R, Pestell RG. Cyclin D1 genetic heterozygosity regulates colonic epithelial cell differentiation and tumor number in ApcMin mice. Mol Cell Biol. 2004;24(17):7598–611.
9. Nakayama J-I, Rice JC, Strahl BD, Allis CD, Grewal SIS. Role of histone H3 lysine 9 methylation in heterochromatin assembly and epigenetic gene silencing. Science. 2001;292:110–3.
10. Neumeister P, Albanese C, Balent B, Greally J, Pestell RG. Senescence and epigenetic dysregulation in cancer. Int J Biochem Cell Biol. 2002;34(11):1475–90.
11. Reutens AT, Fu M, Wang C, Albanese C, McPhaul MJ, Sun Z, Balk SP, Janne OA, Palvimo JJ, Pestell RG. Cyclin D1 binds the androgen receptor and regulates hormone-dependent signaling in a p300/CBP-associated factor (P/CAF)-dependent manner. Mol Endocrinol. 2001;15(5):797–811.
12. Fu M, Wang C, Rao M, Wu X, Bouras T, Zhang X, Li Z, Jiao X, Yang J, Li A, Perkins ND, Thimmapaya B, Kung AL, Munoz A, Giordano A, Lisanti MP, Pestell RG. Cyclin D1 represses p300 transactivation through a cyclin-dependent kinase-independent mechanism. J Biol Chem. 2005;280(33):29728–42.

13. Di Sante G, Di Rocco A, Pupo C, Casimiro MC, Pestell RG. Hormone-induced DNA damage response and repair mediated by cyclin D1 in breast and prostate cancer. Oncotarget. 2017; https://doi.org/10.18632/oncotarget.19413.
14. Bienvenu F, Barre B, Giraud S, Avril S, Coqueret O. Transcriptional regulation by a DNA-associated form of cyclin D1. Mol Biol Cell. 2005;16(4):1850–8.
15. Cicatiello L, Addeo R, Sasso A, Altucci L, Petrizzi VB, Borgo R, Cancemi M, Caporali S, Caristi S, Scafoglio C, Teti D, Bresciani F, Perillo B, Weisz A. Estrogens and progesterone promote persistent CCND1 gene activation during G1 by inducing transcriptional derepression via c-Jun/c-Fos/estrogen receptor (progesterone receptor) complex assembly to a distal regulatory element and recruitment of cyclin D1 to its own gene promoter. Mol Cell Biol. 2004;24(16):7260–74.
16. Jirawatnotai S, Hu Y, Michowski W, Elias JE, Becks L, Bienvenu F, Zagozdzon A, Goswami T, Wang YE, Clark AB, Kunkel TA, van Harn T, Xia B, Correll M, Quackenbush J, Livingston DM, Gygi SP, Sicinski P. A function for cyclin D1 in DNA repair uncovered by protein interactome analyses in human cancers. Nature. 2011;474(7350):230–4.
17. Casimiro MC, Crosariol M, Loro E, Ertel A, Yu Z, Dampier W, Saria EA, Papanikolaou A, Stanek TJ, Li Z, Wang C, Fortina P, Addya S, Tozeren A, Knudsen ES, Arnold A, Pestell RG. ChIP sequencing of cyclin D1 reveals a transcriptional role in chromosomal instability in mice. J Clin Invest. 2012;122(3):833–43.
18. Lee YM, Sicinski P. Targeting cyclins and cyclin-dependent kinases in cancer: lessons from mice, hopes for therapeutic applications in human. Cell Cycle. 2006;5(18):2110–4.
19. Malumbres M, Barbacid M. Cell cycle kinases in cancer. Curr Opin Genet Dev. 2007;17(1):60–5.
20. Sherr CJ, Roberts JM. Living with or without cyclins and cyclin-dependent kinases. Genes Dev. 2004;18(22):2699–711.
21. Hunter T, Pines J. Cyclins and cancer II: cyclin D and CDK inhibitors come of age. Cell. 1994;79:573–82.
22. Weinberg RA. The retinoblastoma protein and cell cycle control. Cell. 1995;81:323–30.
23. Sherr CJ. Cancer cell cycles. Science. 1996;274:1672–7.
24. Rosenthal ET, Hunt T, Ruderman JV. Selective translation of mRNA controls the pattern of protein synthesis during early development of the surf clam. Spisula solidissima Cell. 1980;20(2):487–94.
25. Motokura T, Bloom T, Kim HG, Juppner H, Ruderman JV, Kronenberg HM, Arnold A. A novel cyclin encoded by a bcl1-linked candidate oncogene. Nature. 1991;350(6318):512–5.
26. Xiong Y, Connolly T, Futcher B, Beach D. Human D-type cyclin. Cell. 1991;65(4):691–9.
27. Sherr CJ, Beach D, Shapiro GI. Targeting CDK4 and CDK6: from discovery to therapy. Cancer Discov. 2015;6(4):353.
28. Casimiro MC, Velasco-Velazquez M, Aguirre-Alvarado C, Pestell RG. Overview of cyclins D1 function in cancer and the CDK inhibitor landscape: past and present. Expert Opin Investig Drugs. 2014;23(3):295–304.
29. Sherr CJ. Mammalian G1 cyclins. Cell. 1993;73:1059–65.
30. Dowdy SF, Hinds PW, Louie K, Reed SI, Arnold A, Weinberg RA. Physical interaction of the retinoblastoma protein with human D cyclins. Cell. 1993;73:499–511.
31. Ewen ME, Sluss HK, Sherr CJ, Matsushime H, Kato J, Livingston DM. Functional interactions of the retinoblastoma protein with mammalian D-type cyclins. Cell. 1993;73(3):487–97.
32. Baldin V, Lukas J, Marcote MJ, Pagano M, Draetta G. Cyclin D1 is a nuclear protein required for cell cycle progression in G1. Genes Dev. 1993;7:812–21.
33. Quelle DE, Ashmun RA, Shurtleff SA, Kato J-Y, Bar-Sagi D, Roussel MF, Sherr CJ. Overexpression of mouse D-type cyclins accelerates G1 phase in rodent fibroblasts. Genes Dev. 1993;7(8):1559–71.
34. Marampon F, Casimiro MC, Fu M, Powell MJ, Popov VM, Lindsay J, Zani BM, Ciccarelli C, Watanabe G, Lee RJ, Pestell RG. Mol Biol Cell. 2008;19(6):2566–78. https://doi.org/10.1091/mbc.E06-12-1110. Epub 2008 Mar 26. PMID:18367547

35. Lukas J, Bartkova J, Bartek J. Convergence of mitogenic signalling cascades from diverse classes of receptors at the cyclin D-cyclin-dependent kinase-pRb-controlled G1 checkpoint. Mol Cell Biol. 1996;16(12):6917–25.
36. Jiang W, Kahn SM, Zhou P, Zhang Y-J, Cacace AM, Infante AS, Doi S, Santella RM, Weinstein IB. Overexpression of cyclin D1 in rat fibroblasts causes abnormalities in growth control, cell cycle progression and gene expression. Oncogene. 1993;8(12):3447–57.
37. Resnitzky D, Gossen M, Bujard H, Reed SI. Acceleration of the G1/S phase transition by expression of cyclins D1 and E with an inducible system. Mol Cell Biol. 1994;14(3):1669–79.
38. Matsushime H, Ewen ME, Strom DK, Kato J-Y, Hanks SK, Roussel MF, Sherr CJ. Identification and properties of an atypical catalytic subunit (p34^{PSK-J3}/cdk4) for mammalian D type G1 cyclins. Cell. 1992;71:323–34.
39. Meyerson M, Harlow E. Identification of G1 kinase activity for cdk6, a novel cyclin D partner. Mol Cell Biol. 1994;14(3):2077–86.
40. Kato J-Y, Matsushime H, Hiebert SW, Ewen ME, Sherr CJ. Direct binding of cyclin D to the retinoblastoma gene product (pRb) and pRb phosphorylation by the cyclin D-dependent kinase CDK4. Genes Dev. 1993;7:331–42.
41. Lees JA, Buchkovich KJ, Marshak DR, Anderson CW, Harlow E. The retinoblastoma protein is phosphorylated on multiple sites by human cdc2. EMBO J. 1991;10:4279–90.
42. Kitagawa M, Higashi H, Jung H-K, Suzuki-Takahashi I, Ikeda M, Tamai K, Kato J-Y, Segawa K, Yoshida E, Nishimura S, Taya Y. The consensus motif for phosphorylation by cyclin D1-Cdk4 is different from that for phosphorylation by cyclin a/E-Cdk2. EMBO J. 1996;15(24):7060–9.
43. Lee RJ, Albanese C, Fu M, D'Amico M, Lin B, Watanabe G, Haines GK 3rd, Siegel PM, Hung MC, Yarden Y, Horowitz JM, Muller WJ, Pestell RG. Cyclin D1 is required for transformation by activated Neu and is induced through an E2F-dependent signaling pathway. Mol Cell Biol. 2000;20(2):672–83.
44. Yu Q, Geng Y, Sicinski P. Specific protection against breast cancers by cyclin D1 ablation. Nature. 2001;411(6841):1017–21.
45. Robles AI, Rodriguez-Puebla ML, Glick AB, Trempus C, Hansen L, Sicinski P, Tennant RW, Weinberg RA, Yuspa SH, Conti CJ. Reduced skin tumor development in cyclin D1-deficient mice highlights the oncogenic ras pathway in vivo. Genes Dev. 1998;12(16):2469–74.
46. Ganter B, Fu S, Lipsick JS. D-type cyclins repress transcriptional activation by the v-Myb but not the c-Myb DNA-binding domain. EMBO J. 1998;17(1):255–68.
47. Hirai H, Sherr CJ. Interaction of D-type cyclins with a novel myb-like transcription factor, DMP1. Mol Cell Biol. 1996;16(11):6457–67.
48. Inoue K, Sherr CJ. Gene expression and cell cycle arrest mediated by transcription factor DMP1 is antagonized by D-type cyclins through a cyclin- dependent-kinase-independent mechanism. Mol Cell Biol. 1998;18(3):1590–600.
49. Ratineau C, Petry MW, Mutoh H, Leiter AB. Cyclin D1 represses the basic helix-loop-helix transcription factor, BETA2/NeuroD. J Biol Chem. 2002;277(11):8847–53.
50. Horstmann S, Ferrari S, Klempnauer KH. Regulation of B-Myb activity by cyclin D1. Oncogene. 2000;19(2):298–306.
51. Skapek SX, Rhee J, Kim PS, Novitch BG, Lassar AB. Cyclin-mediated inhibition of muscle gene expression via a mechanism that is independent of pRB hyperphosphorylation. Mol Cell Biol. 1996;16(12):7043–53.
52. Rao SS, Chu C, Kohtz DS. Ectopic expression of cyclin D1 prevents activation of gene transcription by myogenic basic helix-loop-helix regulators. Mol Cell Biol. 1994;14(8):5259–67.
53. Bienvenu F, Gascan H, Coqueret O. Cyclin D1 represses STAT3 activation through a Cdk4-independent mechanism. J Biol Chem. 2001;276(20):16840–7.
54. Pestell RG. New roles of cyclin D1. Am J Pathol. 2013;183(1):3–9.
55. Lin HM, Zhao L, Cheng SY. Cyclin D1 is a ligand-independent co-repressor for thyroid hormone receptors. J Biol Chem. 2002;277(32):28733–41.
56. Lamb J, Ladha MH, McMahon C, Sutherland RL, Ewen ME. Regulation of the functional interaction between Cyclin D1 and the estrogen receptor. Mol Cell Biol. 2000;20:8667–75.

57. Wang C, Pattabiraman N, Zhou JN, Fu M, Sakamaki T, Albanese C, Li Z, Wu K, Hulit J, Neumeister P, Novikoff PM, Brownlee M, Scherer PE, Jones JG, Whitney KD, Donehower LA, Harris EL, Rohan T, Johns DC, Pestell RG. Cyclin D1 repression of peroxisome proliferator-activated receptor gamma expression and transactivation. Mol Cel Biol. 2003;23(17):6159–73.

58. Zwijsen RM, Wientjens E, Klompmaker R, van der Sman J, Bernards R, Michalides RJ. CDK-independent activation of estrogen receptor by cyclin D1. Cell. 1997;88(3):405–15.

59. Neuman E, Ladha MH, Lin N, Upton TM, Miller SJ, DiRenzo J, Pestell RG, Hinds PW, Dowdy SF, Brown M, Ewen ME. Cyclin D1 stimulation of estrogen receptor transcriptional activity independent of cdk4. Mol Cell Biol. 1997;17(9):5338–47.

60. Reutens A, Watanabe G, Albanese C, Pestell RG. Cyclin D1 binds activating mutants of the androgen receptor. US 80th Endocrine Society Meeting, New Orleans. 1998. Abstract P1–528.

61. Knudsen KE, Cavenee WK, Arden KC. D-type cyclins complex with the androgen receptor and inhibit its transcriptional transactivation ability. Cancer Res. 1999;59(10):2297–301.

62. Gu W, Schneider JW, Condorelli G, Kaushal S, Mahdavi V, Nadal-Ginard B. Interaction of myogenic factors and the retinoblastoma protein mediates muscle cell commitment and differentiation. Cell. 1993;72:309–24.

63. Zwijsen RM, Buckle RS, Hijmans EM, Loomans CJ, Bernards R. Ligand-independent recruitment of steroid receptor coactivators to estrogen receptor by cyclin D1. Genes Dev. 1998;12(22):3488–98.

64. Bhalla K, Liu WJ, Thompson K, Anders L, Devarakonda S, Dewi R, Buckley S, Hwang BJ, Polster B, Dorsey SG, Sun Y, Sicinski P, Girnun GD. Cyclin D1 represses gluconeogenesis via inhibition of the transcriptional coactivator PGC1alpha. Diabetes. 2014;63(10):3266–78.

65. Wang C, Li Z, Lu Y, Du R, Katiyar S, Yang J, Fu M, Leader JE, Quong A, Novikoff PM, Pestell RG. Cyclin D1 repression of nuclear respiratory factor 1 integrates nuclear DNA synthesis and mitochondrial function. Proc Natl Acad Sci U S A. 2006;103(31):11567–72.

66. Lamb J, Ramaswamy S, Ford HL, Contreras B, Martinez RV, Kittrell FS, Zahnow CA, Patterson N, Golub TR, Ewen ME. A mechanism of Cyclin D1 action encoded in the patterns of gene expression in human cancer. Cell. 2003;114:323–34.

67. Yamak A, Latinkic BV, Dali R, Temsah R, Nemer M. Cyclin D2 is a GATA4 cofactor in cardiogenesis. Proc Natl Acad Sci U S A. 2014;111(4):1415–20.

68. Chun T, Rho SB, Byun HJ, Lee JY, Kong G. The polycomb group gene product Mel-18 interacts with cyclin D2 and modulates its activity. FEBS Lett. 2005;579(24):5275–80.

69. Peterson LF, Boyapati A, Ranganathan V, Iwama A, Tenen DG, Tsai S, Zhang DE. The hematopoietic transcription factor AML1 (RUNX1) is negatively regulated by the cell cycle protein cyclin D3. Mol Cell Biol. 2005;25(23):10205–19.

70. Sarruf DA, Iankova I, Abella A, Assou S, Miard S, Fajas L. Cyclin D3 promotes adipogenesis through activation of peroxisome proliferator-activated receptor gamma. Mol Cell Biol. 2005;25(22):9985–95.

71. Wang GL, Shi X, Salisbury E, Sun Y, Albrecht JH, Smith RG, Timchenko NA. Cyclin D3 maintains growth-inhibitory activity of C/EBPalpha by stabilizing C/EBPalpha-cdk2 and C/EBPalpha-Brm complexes. Mol Cell Biol. 2006;26(7):2570–82.

72. Liu W, Sun M, Jiang J, Shen X, Sun Q, Liu W, Shen H, Gu J. Cyclin D3 interacts with human activating transcription factor 5 and potentiates its transcription activity. Biochem Biophys Res Commun. 2004;321(4):954–60.

73. Jian Y, Yan J, Wang H, Chen C, Sun M, Jiang J, Lu J, Yang Y, Gu J. Cyclin D3 interacts with vitamin D receptor and regulates its transcription activity. Biochem Biophys Res Commun. 2005;335(3):739–48.

74. Zong H, Chi Y, Wang Y, Yang Y, Zhang L, Chen H, Jiang J, Li Z, Hong Y, Wang H, Yun X, Gu J. Cyclin D3/CDK11p58 complex is involved in the repression of androgen receptor. Mol Cell Biol. 2007;27(20):7125–42.

75. Kim Y, Kim J, Jang SW, Ko J. The role of sLZIP in cyclin D3-mediated negative regula-
    tion of androgen receptor transactivation and its involvement in prostate cancer. Oncogene.
    2015;34(2):226–36.
76. Olshavsky NA, Groh EM, Comstock CE, Morey LM, Wang Y, Revelo MP, Burd C, Meller
    J, Knudsen KE. Cyclin D3 action in androgen receptor regulation and prostate cancer.
    Oncogene. 2008;27(22):3111–21.
77. Reutens A, Watanabe G, Albanese C, McPhaul MJ, Balk SP, Pestell RG. Cyclin D1 binds
    activating mutants of the androgen receptor. US Endocrine Society Meeting, New Orleans,
    Louisiana. Bethesda: Endocrine Society Press; 1998. Abstract number P1-528, p. 228
78. McMahon C, Suthiphongchai T, DiRenzo J, Ewen ME. P/CAF associates with cyclin
    D1 and potentiates its activation of the estrogen receptor. Proc Natl Acad Sci U S A.
    1999;96(10):5382–7.
79. Lee Y, Dominy JE, Choi YJ, Jurczak M, Tolliday N, Camporez JP, Chim H, Lim JH, Ruan HB,
    Yang X, Vazquez F, Sicinski P, Shulman GI, Puigserver P. Cyclin D1-Cdk4 controls glucose
    metabolism independently of cell cycle progression. Nature. 2014;510(7506):547–51.
80. Wang C, Fan S, Li Z, Fu M, Rao M, Ma Y, Lisanti MP, Albanese C, Katzenellenbogen BS,
    Kushner PJ, Weber B, Rosen EM, Pestell RG. Cyclin D1 antagonizes BRCA1 repression of
    estrogen receptor alpha activity. Cancer Res. 2005;65(15):6557–67.
81. Adnane J, Shao Z, Robbins PD. Cyclin D1 associates with the TBP-associated factor
    TAF(II)250 to regulate Sp1-mediated transcription. Oncogene. 1999;18(1):239–47.
82. Sternglanz R, Schindelin H. Structure and mechanism of action of the histone acetyl-
    transferase Gcn5 and similarity to other N-acetyltransferases. Proc Natl Acad Sci U S A.
    1999;96(16):8807–8.
83. Trievel RC, Rojas JR, Sterner DE, Venkataramani RN, Wang L, Zhou J, Allis CD, Berger SL,
    Marmorstein R. Crystal structure and mechanism of histone acetylation of the yeast GCN5
    transcriptional coactivator. Proc Natl Acad Sci U S A. 1999;96(16):8931–6.
84. Clements A, Rojas JR, Trievel RC, Wang L, Berger SL, Marmorstein R. Crystal structure of
    the histone acetyltransferase domain of the human PCAF transcriptional regulator bound to
    coenzyme a. EMBO J. 1999;18(13):3521–32.
85. Mizzen CA, Allis CD. Linking histone acetylation to transcriptional regulation. Cell Mol Life
    Sci. 1998;54(1):6–20.
86. Dancy BM, Cole PA. Protein lysine acetylation by p300/CBP. Chem Rev. 2015;115(6):2419–52.
87. Savage KI, Harkin DP. BRCA1, a 'complex' protein involved in the maintenance of genomic
    stability. FEBS J. 2015;282(4):630–46.
88. Miki Y, Swensen J, Shattuck-Eidens D, Futreal PA, Harshman K, Tavtigian S, Liu Q,
    Cochran C, Bennett LM, Ding W, et al. A strong candidate for the breast and ovarian cancer
    susceptibility gene BRCA1. Science. 1994;266(5182):66–71.
89. Monteiro AN, August A, Hanafusa H. Evidence for a transcriptional activation function of
    BRCA1 C-terminal region. Proc Natl Acad Sci U S A. 1996;93(24):13595–9.
90. Yarden RI, Brody LC. BRCA1 interacts with components of the histone deacetylase complex.
    Proc Natl Acad Sci U S A. 1999;96:4983–8.
91. Fan S, Wang J, Yuan R, Ma Y, Meng Q, Erdos MR, Pestell RG, Yuan F, Auborn KJ, Goldberg
    ID, Rosen EM. BRCA1 inhibition of estrogen receptor signaling in transfected cells. Science.
    1999;284(5418):1354–6.
92. Fan S, Ma YX, Wang C, Yuan RQ, Meng Q, Wang JA, Erdos M, Goldberg ID, Webb P,
    Kushner PJ, Pestell RG, Rosen EM. Role of direct interaction in BRCA1 inhibition of estrogen
    receptor activity. Oncogene. 2001;20(1):77–87.
93. Fan S, Ma YX, Wang C, Yuan RQ, Meng Q, Wang JA, Erdos M, Goldberg ID, Webb P,
    Kushner PJ, Pestell RG, Rosen EM. p300 modulates the BRCA1 inhibition of estrogen receptor
    activity. Cancer Res. 2002;62(1):141–51.
94. Puigserver P. Tissue-specific regulation of metabolic pathways through the transcriptional
    coactivator PGC1-alpha. Int J Obes. 2005;29(Suppl 1):S5–9.

95. Wu Z, Puigserver P, Andersson U, Zhang C, Adelmant G, Mootha V, Troy A, Cinti S, Lowell B, Scarpulla RC, Spiegelman BM. Mechanisms controlling mitochondrial biogenesis and respiration through the thermogenic coactivator PGC-1. Cell. 1999;98(1):115–24.

96. Sakamaki T, Casimiro MC, Ju X, Quong AA, Katiyar S, Liu M, Jiao X, Li A, Zhang X, Lu Y, Wang C, Byers S, Nicholson R, Link T, Shemluck M, Yang J, Fricke ST, Novikoff PM, Papanikolaou A, Arnold A, Albanese C, Pestell R. Cyclin D1 determines mitochondrial function in vivo. Mol Cell Biol. 2006;26(14):5449–69.

97. Hanse EA, Mashek DG, Becker JR, Solmonson AD, Mullany LK, Mashek MT, Towle HC, Chau AT, Albrecht JH. Cyclin D1 inhibits hepatic lipogenesis via repression of carbohydrate response element binding protein and hepatocyte nuclear factor 4alpha. Cell Cycle. 2012;11(14):2681–90.

98. Comstock CE, Augello MA, Schiewer MJ, Karch J, Burd CJ, Ertel A, Knudsen ES, Jessen WJ, Aronow BJ, Knudsen KE. Cyclin D1 is a selective modifier of androgen-dependent signaling and androgen receptor function. J Biol Chem. 2011;286(10):8117–27.

99. Ju X, Casimiro MC, Gormley M, Meng H, Jiao X, Katiyar S, Crosariol M, Chen K, Wang M, Quong AA, Lisanti MP, Ertel A, Pestell RG. Identification of a cyclin D1 network in prostate cancer that antagonizes epithelial-mesenchymal restraint. Cancer Res. 2014;74(2):508–19.

100. Hisatake K, Hasegawa S, Takada R, Nakatani Y, Horikoshi M, Roeder RG. The p250 subunit of native TATA box-binding factor TFIID is the cell-cycle regulatory protein CCG1. Nature. 1993;362(6416):179–81.

101. Sekiguchi T, Miyata T, Nishimoto T. Molecular cloning of the cDNA of human X chromosomal gene (CCG1) which complements the temperature-sensitive G1 mutants, tsBN462 and ts13, of the BHK cell line. EMBO J. 1988;7(6):1683–7.

102. Sekiguchi T, Nohiro Y, Nakamura Y, Hisamoto N, Nishimoto T. The human CCG1 gene, essential for progression of the G1 phase, encodes a 210-kilodalton nuclear DNA-binding protein. Mol Cell Biol. 1991;11(6):3317–25.

103. Suzuki-Yagawa Y, Guermah M, Roeder RG. The ts13 mutation in the TAF$_{II}$250 subunit (CCG1) of TFIID directly affects transcription of D-type cyclin genes in cells arrested in G1 at the nonpermissive temperature. Mol Cell Biol. 1997;17(6):3284–94.

104. Kim SJ, Onwuta US, Lee YI, Li R, Botchan MR, Robbins PD. The retinoblastoma gene product regulates Sp1-mediated transcription. Mol Cell Biol. 1992;12(6):2455–63.

105. Shao Z, Robbins PD. Differential regulation of E2F and Sp1-mediated transcription by G1 cyclins. Oncogene. 1995;10(2):221–8.

106. Tessarz P, Kouzarides T. Histone core modifications regulating nucleosome structure and dynamics. Nat Rev Mol Cell Biol. 2014;15(11):703–8.

107. Bannister AJ, Kouzarides T. Regulation of chromatin by histone modifications. Cell Res. 2011;21(3):381–95.

108. Liu WD, Wang HW, Muguira M, Breslin MB, Lan MS. INSM1 functions as a transcriptional repressor of the neuroD/beta2 gene through the recruitment of cyclin D1 and histone deacetylases. Biochem J. 2006;397(1):169–77.

109. Wang C, Fu M, Pestell RG. Histone acetylation/deacetylation as a regulator of cell cycle gene expression. Methods Mol Biol. 2004;241:207–16.

110. Al-Sady B, Madhani HD, Narlikar GJ. Division of labor between the chromodomains of HP1 and Suv39 methylase enables coordination of heterochromatin spread. Mol Cell. 2013;51(1):80–91.

111. Li Y, Diehl JA. PRMT5-dependent p53 escape in tumorigenesis. Oncoscience. 2015;2(8):700–2.

112. Aggarwal P, Vaites LP, Kim JK, Mellert H, Gurung B, Nakagawa H, Herlyn M, Hua X, Rustgi AK, McMahon SB, Diehl JA. Nuclear cyclin D1/CDK4 kinase regulates CUL4 expression and triggers neoplastic growth via activation of the PRMT5 methyltransferase. Cancer Cell. 2010;18(4):329–40.

113. Li Y, Chitnis N, Nakagawa H, Kita Y, Natsugoe S, Yang Y, Li Z, Wasik M, Klein-Szanto AJ, Rustgi AK, Diehl JA. PRMT5 is required for lymphomagenesis triggered by multiple oncogenic drivers. Cancer Discov. 2015;5(3):288–303.

114. Zheng S, Moehlenbrink J, Lu YC, Zalmas LP, Sagum CA, Carr S, McGouran JF, Alexander L, Fedorov O, Munro S, Kessler B, Bedford MT, Yu Q, La Thangue NB. Arginine methylation-dependent reader-writer interplay governs growth control by E2F-1. Mol Cell. 2013;52(1):37–51.
115. Bienvenu F, Jirawatnotai S, Elias JE, Meyer CA, Mizeracka K, Marson A, Frampton GM, Cole MF, Odom DT, Odajima J, Geng Y, Zagozdzon A, Jecrois M, Young RA, Liu XS, Cepko CL, Gygi SP, Sicinski P. Transcriptional role of cyclin D1 in development revealed by a genetic-proteomic screen. Nature. 2010;463(7279):374–8.
116. Chellappan SP, Hiebert S, Mudryj M, Horowitz JM, Nevins JR. The E2F transcription factor is a cellular target for the RB protein. Cell. 1991;65(6):1053–61.
117. Raychaudhuri P, Rooney R, Nevins JR. Identification of an E1A-inducible cellular factor that interacts with regulatory sequences within the adenovirus E4 promoter. EMBO J. 1987;6(13):4073–81.
118. Yee AS, Raychaudhuri P, Jakoi L, Nevins JR. The adenovirus-inducible factor E2F stimulates transcription after specific DNA binding. Mol Cell Biol. 1989;9(2):578–85.
119. Partridge JF, La Thangue NB. A developmentally regulated and tissue-dependent transcription factor complexes with the retinoblastoma gene product. EMBO J. 1991;10(12):3819–27.
120. Bandara LR, Adamczewski JP, Hunt T, La Thangue NB. Cyclin a and the retinoblastoma gene product complex with a common transcription factor. Nature. 1991;352(6332):249–51.
121. Watanabe G, Albanese C, Lee RJ, Reutens A, Vairo G, Henglein B, Pestell RG. Inhibition of cyclin D1 kinase activity is associated with E2F-mediated inhibition of cyclin D1 promoter activity through E2F and Sp1. Mol Cell Biol. 1998;18(6):3212–22.
122. Malumbres M, Barbacid M. Cell cycle, CDKs and cancer: a changing paradigm. Nat Rev Cancer. 2009;9(3):153–66.
123. Lengauer C, Kinzler KW, Vogelstein B. Genetic instabilities in human cancers. Nature. 1998;396(6712):643–9.
124. Thompson SL, Bakhoum SF, Compton DA. Mechanisms of chromosomal instability. Curr Biol. 2010;20(6):R285–95.
125. Gollin SM. Mechanisms leading to chromosomal instability. Semin Cancer Biol. 2005;15(1):33–42.
126. Draviam VM, Xie S, Sorger PK. Chromosome segregation and genomic stability. Curr Opin Genet Dev. 2004;14(2):120–5.
127. Spruck CH, Won KA, Reed SI. Deregulated cyclin E induces chromosome instability. Nature. 1999;401(6750):297–300.
128. Carter SL, Eklund AC, Kohane IS, Harris LN, Szallasi Z. A signature of chromosomal instability inferred from gene expression profiles predicts clinical outcome in multiple human cancers. Nat Genet. 2006;38(9):1043–8.
129. Casimiro MC, Di Sante G, Crosariol M, Loro E, Dampier W, Ertel A, Yu Z, Saria EA, Papanikolaou A, Li Z, Wang C, Addya S, Lisanti MP, Fortina P, Cardiff RD, Tozeren A, Knudsen ES, Arnold A, Pestell RG. Kinase-independent role of cyclin D1 in chromosomal instability and mammary tumorigenesis. Oncotarget. 2015;6(11):8525–38.
130. Casimiro MC, Arnold A, Pestell RG. Kinase independent oncogenic cyclin D1. Aging (Albany NY). 2015;7(7):455–6.
131. Inoue K, Zindy F, Randle DH, Rehg JE, Sherr CJ. Dmp1 is haplo-insufficient for tumor suppression and modifies the frequencies of Arf and p53 mutations in Myc-induced lymphomas. Genes Dev. 2001;15(15):2934–9.
132. Zhu S, Mott RT, Fry EA, Taneja P, Kulik G, Sui G, Inoue K. Cooperation between Dmp1 loss and cyclin D1 overexpression in breast cancer. Am J Pathol. 2013;183(4):1339–50.
133. Maglic D, Stovall DB, Cline JM, Fry EA, Mallakin A, Taneja P, Caudell DL, Willingham MC, Sui G, Inoue K. DMP1beta, a splice isoform of the tumour suppressor DMP1 locus, induces proliferation and progression of breast cancer. J Pathol. 2015;236(1):90–102.
134. Inoue K, Fry EA. Aberrant splicing of the DMP1-ARF-MDM2-p53 pathway in cancer. Int J Cancer. 2016;139(1):33.

135. Liu Q, Boudot A, Ni J, Hennessey T, Beauparlant SL, Rajabi HN, Zahnow C, Ewen ME. Cyclin D1 and C/EBPbeta LAP1 operate in a common pathway to promote mammary epithelial cell differentiation. Mol Cell Biol. 2014;34(16):3168–79.
136. Casimiro MC, Wang C, Li Z, Di Sante G, Willmart NE, Addya S, Chen L, Liu Y, Lisanti MP, Pestell RG. Cyclin D1 determines estrogen signaling in the mammary gland in vivo. Mol Endocrinol. 2013;27(9):1415–28.
137. Liu MM, Albanese C, Anderson CM, Hilty K, Webb P, Uht RM, Price RH Jr, Pestell RG, Kushner PJ. Opposing action of estrogen receptors alpha and beta on cyclin D1 gene expression. J Biol Chem. 2002;277(27):24353–60.
138. Levin ER. Extranuclear estrogen receptor's roles in physiology: lessons from mouse models. Am J Physiol Endocrinol Metab. 2014;307(2):E133–40.
139. Li Z, Chen K, Jiao X, Wang C, Willmarth NE, Casimiro MC, Li W, Ju X, Kim SH, Lisanti MP, Katzenellenbogen JA, Pestell RG. Cyclin D1 integrates estrogen-mediated DNA damage repair signaling. Cancer Res. 2014;74(14):3959–70.
140. Chen Y, Martinez LA, LaCava M, Coghlan L, Conti CJ. Increased cell growth and tumorigenicity in human prostate LNCaP cells by overexpression to cyclin D1. Oncogene. 1998;16(15):1913–20.
141. Casimiro MC, Di Sante G, Ju X, Li Z, Chen K, Crosariol M, Yaman I, Gormley M, Meng H, Lisanti MP, Pestell RG. Cyclin D1 promotes androgen-dependent DNA damage repair in prostate cancer cells. Cancer Res. 2016;76(2):329–38.
142. Marampon F, Gravina GL, Ju X, Vetuschi A, Sferra R, Casimiro MC, Pompili S, Festuccia C, Colapietro A, Gaudio E, Di Cesare E, Tombolini V, Pestell RG. Cyclin D1 silencing suppresses tumorigenicity, impairs DNA double strand break repair and thus radiosensitizes androgenindependent prostate cancer cells to DNA damage. Oncotarget. 2015;7(5):5383.
143. Ding Z, Wu CJ, Chu GC, Xiao Y, Ho D, Zhang J, Perry SR, Labrot ES, Wu X, Lis R, Hoshida Y, Hiller D, Hu B, Jiang S, Zheng H, Stegh AH, Scott KL, Signoretti S, Bardeesy N, Wang YA, Hill DE, Golub TR, Stampfer MJ, Wong WH, Loda M, Mucci L, Chin L, DePinho RA. SMAD4-dependent barrier constrains prostate cancer growth and metastatic progression. Nature. 2011;470(7333):269–73.
144. Zhu Z, Zhang H, Lang F, Liu G, Gao D, Li B, Liu Y. Pin1 promotes prostate cancer cell proliferation and migration through activation of Wnt/beta-catenin signaling. Clin Transl Oncol. 2015;18(8):792.
145. Shukla S, Shankar E, Fu P, MacLennan GT, Gupta S. Suppression of NF-kappaB and NF-kappaB-regulated gene expression by Apigenin through IkappaBalpha and IKK pathway in TRAMP mice. PLoS One. 2015;10(9):e0138710.
146. Kumar AP, Bhaskaran S, Ganapathy M, Crosby K, Davis MD, Kochunov P, Schoolfield J, Yeh IT, Troyer DA, Ghosh R. Akt/cAMP-responsive element binding protein/cyclin D1 network: a novel target for prostate cancer inhibition in transgenic adenocarcinoma of mouse prostate model mediated by Nexrutine, a Phellodendron Amurense bark extract. Clin Cancer Res. 2007;13(9):2784–94.
147. Burd CJ, Petre CE, Morey LM, Wang Y, Revelo MP, Haiman CA, Lu S, Fenoglio-Preiser CM, Li J, Knudsen ES, Wong J, Knudsen KE. Cyclin D1b variant influences prostate cancer growth through aberrant androgen receptor regulation. Proc Natl Acad Sci U S A. 2006;103(7):2190–5.
148. Betticher DC, Thatcher N, Altermatt HJ, Hoban P, Ryder WD, Heighway J. Alternate splicing produces a novel cyclin D1 transcript. Oncogene. 1995;11(5):1005–11.
149. Sawa H, Ohshima TA, Ukita H, Murakami H, Chiba Y, Kamada H, Hara M, Saito I. Alternatively spliced forms of cyclin D1 modulate entry into the cell cycle in an inverse manner. Oncogene. 1998;16:1701–12.
150. Wang L, Habuchi T, Mitsumori K, Li Z, Kamoto T, Kinoshita H, Tsuchiya N, Sato K, Ohyama C, Nakamura A, Ogawa O, Kato T. Increased risk of prostate cancer associated with AA genotype of cyclin D1 gene A870G polymorphism. Int J Cancer. 2003;103(1):116–20.
151. Chen Y, Li T, Yu X, Xu J, Li J, Luo D, Mo Z, Hu Y. The RTK/ERK pathway is associated with prostate cancer risk on the SNP level: a pooled analysis of 41 sets of data from case-control studies. Gene. 2014;534(2):286–97.

152. Zheng M, Wan L, He X, Qi X, Liu F, Zhang DH. Effect of the CCND1 A870G polymorphism on prostate cancer risk: a meta-analysis of 3,820 cases and 3,825 controls. World J Surg Oncol. 2015;13:55.

153. Yu Z, Wang C, Wang M, Li Z, Casimiro MC, Liu M, Wu K, Whittle J, Ju X, Hyslop T, McCue P, Pestell RG. A cyclin D1/microRNA 17/20 regulatory feedback loop in control of breast cancer cell proliferation. J Cell Biol. 2008;182(3):509–17.

154. Yu Z, Willmarth NE, Zhou J, Katiyar S, Wang M, Liu Y, McCue PA, Quong AA, Lisanti MP, Pestell RG. microRNA 17/20 inhibits cellular invasion and tumor metastasis in breast cancer by heterotypic signaling. Proceedings of the National Academy of Sciences USA. 2010;107(18):8231–6.

155. Yu Z, Wang L, Wang C, Ju X, Wang M, Chen K, Loro E, Li Z, Zhang Y, Wu K, Casimiro MC, Gormley M, Ertel A, Fortina P, Chen Y, Tozeren A, Liu Z, Pestell RG. Cyclin D1 induction of Dicer governs microRNA processing and expression in breast cancer. Nat Commun. 2013;4:2812.

156. Moses RE, O'Malley BW. DNA transcription and repair: a confluence. J Biol Chem. 2012;287(28):23266–70.

157. Malovannaya A, Lanz RB, Jung SY, Bulynko Y, Le NT, Chan DW, Ding C, Shi Y, Yucer N, Krenciute G, Kim BJ, Li C, Chen R, Li W, Wang Y, O'Malley BW, Qin J. Analysis of the human endogenous coregulator complexome. Cell. 2011;145(5):787–99.

158. Ju BG, Lunyak VV, Perissi V, Garcia-Bassets I, Rose DW, Glass CK, Rosenfeld MG. A topoisomerase IIbeta-mediated dsDNA break required for regulated transcription. Science. 2006;312(5781):1798–802.

159. Scully R, Anderson SF, Chao DM, Wei W, Ye L, Young RA, Livingston DM, Parvin JD. BRCA1 is a component of the RNA polymerase II holoenzyme. Proc Natl Acad Sci U S A. 1997;94(11):5605–10.

160. Scully R, Chen J, Ochs RL, Keegan K, Hoekstra M, Feunteun J, Livingston DM. Dynamic changes of BRCA1 subnuclear location and phosphorylation state are initiated by DNA damage. Cell. 1997;90:425–35.

161. Gowen LC, Avrutskaya AV, Latour AM, Koller BH, Leadon SA. BRCA1 required for transcription-coupled repair of oxidative DNA damage. Science. 1998;281(5379):1009–12.

162. Ma Y, Fan S, Hu C, Meng Q, Fuqua SA, Pestell RG, Tomita YA, Rosen EM. BRCA1 regulates acetylation and ubiquitination of estrogen receptor-alpha. Mol Endocrinol. 2010;24(1):76–90.

163. Lin HR, Ting NS, Qin J, Lee WH. M phase-specific phosphorylation of BRCA2 by polo-like kinase 1 correlates with the dissociation of the BRCA2-P/CAF complex. J Biol Chem. 2003;278(38):35979–87.

164. Albanese C, D'Amico M, Reutens AT, Fu M, Watanabe G, Lee RJ, Kitsis RN, Henglein B, Avantaggiati M, Somasundaram K, Thimmapaya B, Pestell RG. Activation of the cyclin D1 gene by the E1A-associated protein p300 through AP-1 inhibits cellular apoptosis. J Biol Chem. 1999;274:34186–95.

165. Geng Y, Lee YM, Welcker M, Swanger J, Zagozdzon A, Winer JD, Roberts JM, Kaldis P, Clurman BE, Sicinski P. Kinase-independent function of cyclin E. Mol Cell. 2007;25(1):127–39.

166. Coverley D, Laman H, Laskey RA. Distinct roles for cyclins E and a during DNA replication complex assembly and activation. Nat Cell Biol. 2002;4(7):523–8.

167. Coco Martin JM. A. Balkenende, T. Verschoor, F. Lallemand, and R. Michalides, Cyclin D1 overexpression enhances radiation-induced apoptosis and radiosensitivity in a breast tumor cell line. Cancer Res. 1999;59(5):1134–40.

168. Zhou Q, Fukushima P, DeGraff W, Mitchell JB, Stetler Stevenson M, Ashkenazi A, Steeg PS. Radiation and the Apo2L/TRAIL apoptotic pathway preferentially inhibit the colonization of premalignant human breast cells overexpressing cyclin D1. Cancer Res. 2000;60(10):2611–5.

169. Agami R, Bernards R. Distinct initiation and maintenance mechanisms cooperate to induce G1 cell cycle arrest in response to DNA damage. Cell. 2000;102(1):55–66.

170. Aggarwal P, Lessie MD, Lin DI, Pontano L, Gladden AB, Nuskey B, Goradia A, Wasik MA, Klein-Szanto AJ, Rustgi AK, Bassing CH, Diehl JA. Nuclear accumulation of cyclin D1 during S phase inhibits Cul4-dependent Cdt1 proteolysis and triggers p53-dependent DNA rereplication. Genes Dev. 2007;21(22):2908–22.

171. Li Z, Jiao X, Wang C, Shirley LA, Elsaleh H, Dahl O, Wang M, Soutoglou E, Knudsen ES, Pestell RG. Alternative cyclin d1 splice forms differentially regulate the DNA damage response. Cancer Res. 2010;70(21):8802–11.

172. Muller GA, Wintsche A, Stangner K, Prohaska SJ, Stadler PF, Engeland K. The CHR site: definition and genome-wide identification of a cell cycle transcriptional element. Nucleic Acids Res. 2014;42(16):10331–50.

173. Neumeister P, Pixley FJ, Xiong Y, Xie H, Wu K, Ashton A, Cammer M, Chan A, Symons M, Stanley ER, Pestell RG. Cyclin D1 governs adhesion and motility of macrophages. Mol Biol Cell. 2003;14(5):2005–15.

174. Li Z, Wang C, Jiao X, Lu Y, Fu M, Quong AA, Dye C, Yang J, Dai M, Ju X, Zhang X, Li A, Burbelo P, Stanley ER, Pestell RG. Cyclin D1 regulates cellular migration through the inhibition of thrombospondin 1 and ROCK signaling. Mol Cel Biol. 2006;26(11):4240–56.

175. Li Z, Jiao X, Wang C, Ju X, Lu Y, Yuan L, Lisanti MP, Katiyar S, Pestell RG. Cyclin D1 induction of cellular migration requires p27(KIP1). Cancer Res. 2006;66(20):9986–94.

176. Meng H, Tian L, Zhou J, Li Z, Jiao X, Li WW, Plomann M, Xu Z, Lisanti MP, Wang C, Pestell RG. PACSIN 2 represses cellular migration through direct association with cyclin D1 but not its alternate splice form cyclin D1b. Cell Cycle. 2011;10(1):73–81.

177. Dupont S, Morsut L, Aragona M, Enzo E, Giulitti S, Cordenonsi M, Zanconato F, Le Digabel J, Forcato M, Bicciato S, Elvassore N, Piccolo S. Role of YAP/TAZ in mechanotransduction. Nature. 2011;474(7350):179–83.

178. Janmey PA, Wells RG, Assoian RK, McCulloch CA. From tissue mechanics to transcription factors. Differentiation. 2013;86(3):112–20.

179. Kessels MM, Qualmann B. The syndapin protein family: linking membrane trafficking with the cytoskeleton. J Cell Sci. 2004;117(Pt 15):3077–86.

180. Powers SE, Mandal M, Matsuda S, Miletic AV, Cato MH, Tanaka A, Rickert RC, Koyasu S, Clark MR. Subnuclear cyclin D3 compartments and the coordinated regulation of proliferation and immunoglobulin variable gene repression. J Exp Med. 2012;209(12):2199–213.

181. Kind J, Pagie L, Ortabozkoyun H, Boyle S, de Vries SS, Janssen H, Amendola M, Nolen LD, Bickmore WA, van Steensel B. Single-cell dynamics of genome-nuclear lamina interactions. Cell. 2013;153(1):178–92.

182. Zullo JM, Demarco IA, Pique-Regi R, Gaffney DJ, Epstein CB, Spooner CJ, Luperchio TR, Bernstein BE, Pritchard JK, Reddy KL, Singh H. DNA sequence-dependent compartmentalization and silencing of chromatin at the nuclear lamina. Cell. 2012;149(7):1474–87.

183. Towbin BD, Gonzalez-Aguilera C, Sack R, Gaidatzis D, Kalck V, Meister P, Askjaer P, Gasser SM. Step-wise methylation of histone H3K9 positions heterochromatin at the nuclear periphery. Cell. 2012;150(5):934–47.

184. Prokocimer M, Davidovich M, Nissim-Rafinia M, Wiesel-Motiuk N, Bar DZ, Barkan R, Meshorer E, Gruenbaum Y. Nuclear lamins: key regulators of nuclear structure and activities. J Cell Mol Med. 2009;13(6):1059–85.

Chapter 4
Splice Variants and Phosphorylated Isoforms of Cyclin D1 in Tumorigenesis

J. Alan Diehl and Karen E. Knudsen

Abstract Mammalian cells encode three highly homologous D-type cyclins (D1, D2, D3) that associate in a tissue-specific manner with either CDK4 or CDK6 to form an active protein kinase. The D-type cyclin/CDK kinase coordinates G1 progression in response to growth factor signaling. The cyclin D/CDK4 or CDK6 kinase is the first cyclin/CDK complex to be activated in mammalian cells during G1/S transition. While the three D-type cyclins are almost indistinguishable biochemically, cyclin D1 is the most frequently overexpressed or dysregulated in human cancer. The nature of this selectivity remains to be fully understood. While overexpression of cyclin D1 and its ensuing accumulation in tumor cell nuclei frequently result from chromosomal translocations or gene amplification events, such events do not represent the sole source of cyclin D1 dysregulation in human cancer. In this chapter, we discuss the role of posttranscriptional regulation of cyclin D1, the contribution of dysregulation of such regulatory events to human cancer, and the potential therapeutic opportunities this knowledge may afford.

Keywords Cyclin D1 • Cyclin D1b • GSK3β • Phosphorylation • Alternative splicing • CDK4

4.1 Introduction

The dysregulation of mitogen-dependent signaling pathways is a well-established hallmark of cancer. As a consequence, tumor-derived cells generally exhibit reduced growth factor requirements. Cell cycle regulators, particularly those

J.A. Diehl (✉)
Department of Biochemistry and Molecular Biology, Hollings Cancer Center,
Medical University of South Carolina, Charleston, SC 29425, USA
e-mail: diehl@musc.edu

K.E. Knudsen
Departments of Cancer Biology, Urology, Radiation Oncology, and Medical Oncology,
Sidney Kimmel Cancer Center, Thomas Jefferson University, Philadelphia, PA 19107, USA

© Springer International Publishing AG 2018   91
P.W. Hinds, N.E. Brown (eds.), *D-type Cyclins and Cancer*,
Current Cancer Research, DOI 10.1007/978-3-319-64451-6_4

governing mitogen-dependent G_1 phase progression, are significantly targeted during oncogenesis as their dysregulation provides cells with a definitive prolif-erative advantage. Mammalian cells contain genes encoding three highly homolo-gous D-type cyclins (D1, D2, D3) that associate in a cell- or tissue-specific manner with either CDK4 or CDK6 to form active protein kinases [1–3]. The cyclin D/CDK4 or cyclin D1/CDK6 kinases are the first cyclin/CDK complexes to be acti-vated in mammalian cells. As discussed below, the activation of these complexes represents a key event during the cell cycle in which integration of growth factors and other mitogen-dependent signals occurs [1]. Moreover, the D-type cyclins also hold distinct kinase-dependent and kinase-independent roles in transcrip-tional regulation and DNA repair [4–7]. While the three D-type cyclins are almost indistinguishable biochemically, the gene encoding cyclin D1 is the most fre-quently overexpressed or otherwise dysregulated in human cancers [8]. The nature of this selectivity remains a poorly understood issue. While overexpression of cyclin D1 and its ensuing increased accumulation in tumor cell nuclei frequently result from chromosomal translocations or gene amplification events, as discussed subsequently, such events do not represent the sole source of cyclin D1 dysregula-tion in human cancer.

The gene that encodes cyclin D1, *CCND1*, was first identified at the sites the chromosomal translocation (11p15;q13) on human chromosome 11 [9], in a para-thyroid adenoma, and as the *BCL1* oncogene in a chromosomal translocation observed in mantle B cell lymphoma (MCL) [10]. MCL is an aggressive malig-nancy that accounts for approximately 10% of non-Hodgkin lymphomas. Indeed, the t(11;14)(q13;q32) translocation in MCL, wherein the coding region of cyclin D1 is juxtaposed to the immunoglobulin heavy chain gene, is considered a hall-mark of this disease [11]. As a result of this translocation, high levels of the cyclin D1 protein are found in lymphoid cells, where normally only cyclins D2 and D3 are expressed, potentially contributing to the neoplastic conversion of the afflicted B lymphocytes. Translocations involving t(11;14)(q13;q32) are also associated with 15–20% of multiple myelomas, a malignant tumor arising in germinal centers [12]. On the other hand, translocations involving 6p21, corresponding to the cyclin D3 locus (*CCND3*), are observed in 5% of multiple myelomas [12].

Gene amplification also contributes to overexpression of cyclin D1 in numer-ous cancers. Thus, amplification of 11q13 is observed in several adult cancers, including 30–46% of non-small cell lung cancers [13, 14], 30–50% of head and neck squamous cell carcinomas [15–17], 40% of esophageal squamous cell car-cinomas and adenocarcinomas [18], 25% of pancreatic carcinomas [19], 15% of bladder cancers [20], 49–54% of pituitary adenomas [21, 22], and 13% of breast carcinomas [23, 24]. Although the amplicons are large and contain other genes in addition to *CCND1*, the correlation between gene amplification and high levels of cyclin D1 protein expression, along with the demonstrated ability of cyclin D1 to trigger cancerous phenotypes in genetically modified mice, suggests that *CCND1* dysregulation contributes directly to the malignant phenotype. Importantly, this property is unique to cyclin D1 with regard to other genes within the amplicon.

Since cyclin D1 function is important for growth factor-driven G1/S phase transition, it is reasonable that in cancers in which cyclin D1 levels are normal or even reduced relative to normal cells, compensatory mutations or other alterations in downstream targets of cyclin D1/CDK4 must occur. Downstream targets not only include the retinoblastoma tumor suppressor protein (pRB), a noted cyclin D1/CDK4 substrate, but also negative regulators of the cyclin D1/CDK4 kinase. The net effect of such mutations would be phenotypically similar to dysregulated cyclin D1/CDK4 activity without the need for cyclin D1 overexpression. Consistently, inactivating mutations are found in either p16^{Ink4a} or pRB in a high percentage of human cancers, and these events are typically mutually exclusive with cyclin D1 overexpression [25].

While chromosomal translocations and gene amplification can drive cyclin D1 overexpression, these two events cannot account for the majority of cancers wherein cyclin D1 is overexpressed, suggesting that posttranscriptional and/or posttranslational control of cyclin D1 is of central importance for the maintenance of homeostatic cyclin D1 levels. Over the past decade, significant strides in the elucidation of posttranscriptional control of cyclin D1 have been made. In the following sections, we describe the current understanding of the posttranscriptional mechanisms that regulate cyclin D1 accumulation, with a particular focus on those that are disrupted in human cancer and contribute to neoplastic growth.

4.2 Growth Factor-Dependent Regulation of Cyclin D1

CCND1 expression and cyclin D1 protein translation and assembly with CDK4 are of central importance for mitogen-dependent transition through the restriction point [18, 26], an empirically determined point in the cell cycle wherein cells no longer require growth factor stimulation to complete one round of cell division. Growth factor-mediated regulation of cyclin D1 accumulation and assembly is primarily mediated by Ras-dependent pathways (Fig. 4.1; [27–29]). Thus, cyclin D1 expression and its binding to CDK4 are regulated by the sequential activities of canonical MAPK signaling: RAF kinase, mitogen-activated protein kinase kinases (MEK1 and 2), and the sustained activation of extracellular signal-regulated protein kinases (ERKs; Fig. 4.1) [30–36].

The cyclin D1/CDK4 complexes perform two functions critical for the passage through the restriction point at the late G_1 phase (Fig. 4.2). First, they are involved in the phosphorylation-dependent inactivation of the so-called pocket proteins, namely, the retinoblastoma protein (pRB) and pRB-related family proteins p107 and p130 [37, 38]. pRB, p107, and p130 are considered gatekeepers of the cell cycle that, once activated, antagonize cell division. Cell cycle progression depends upon their phosphorylation, which in turn triggers the release of E2F transcription factor complexes. E2F complexes free from pRB are potent transcriptional activators of genes whose products (such as cyclin E) regulate both the G_1/S transition and S phase [5].

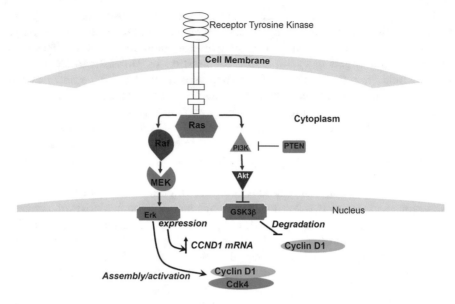

Fig. 4.1 Growth factor regulation of cyclin D1 function and accumulation

Fig. 4.2 Cyclin D1/CDK4 complexes titrate p27 from cyclin E/CDK2 complexes to facilitate E/CDK2 activation

The second key function of cyclin D1/CDK4 complexes involves the stoichiometric titration of members of the CIP/KIP family of CDK inhibitory proteins (Fig. 4.2). This titrating activity in turn facilitates the activation of cyclin E/CDK2 complexes, thereby indirectly regulating entry into the DNA synthetic phase of the cell division cycle [4, 39–44]. It should be noted that binding of CIP/KIP proteins is not an inhibitory event for the cyclin D1/CDK4 kinase complexes. Rather, the incorporation of CIP/KIP proteins facilitates the assembly of the cyclin D1 with CDK4 [45, 46] and also ensures nuclear localization of cyclin D1/CDK4 complexes via inhibition of cyclin D1 nuclear export during the G_1 phase of the cell cycle [47].

4.3 Regulated Phosphorylation of Cyclin D1

Shortly after its identification, further biochemical analyses revealed that cyclin D1 was itself phosphorylated. However, neither the upstream kinase responsible for the phosphorylation of cyclin D1 nor the biochemical role of this posttranslational modification was immediately obvious. Insights into this regulated phosphorylation event initially came from biochemical analysis of the mechanisms regulating cyclin D1 protein stability. Cyclin D1 is a highly labile protein, with a half-life ranging from 20 to 40 min [26]. Work from a number of laboratories revealed that many, if not all, cell cycle regulatory proteins are subject to regulation by a protein degradation factory termed the *u*biquitin *p*roteasome *s*ystem (UPS). At the heart of the UPS is the 26S proteasome, a large multi-subunit complex that harbors three distinct protease activities [27].

A key regulatory aspect of the UPS is the need for the attachment of four or more 8 kDa ubiquitin proteins to any protein destined for UPS-mediated destruction. Ubiquitin chains are recognized by proteins within the 26S proteasome, which in turn unfold the ubiquitylated protein in order to facilitate its transit into the proteolytic core for subsequent degradation. Because many cell cycle regulators are subject to ubiquitin-mediated degradation, early experiments with cyclin D1 initially focused on the assessment of ubiquitin attachment. Indeed, the use of proteasome inhibitors resulted in a reduced rate of cyclin D1 degradation and accumulation of polyubiquitylated cyclin D1 species [26].

Attachment of ubiquitin chains to proteins destined for proteasomal degradation requires the sequential action of three enzymes: the E1-activating enzyme, the E2 ubiquitin-conjugating enzyme, and the E3 ubiquitin ligase. Substrate specificity/selectivity of the UPS generally relies on the E3 ubiquitin ligase. Because it becomes essential that cyclin accumulation is not compromised too early, which would result in cell cycle arrest, degradation generally necessitates a signal in the form of phosphorylation. Phospho-peptide mapping studies revealed a single site of stoichiometric phosphorylation in cyclin D1. Threonine 286, a residue evolutionarily conserved from flies to humans, was identified as the key target for phosphorylation. Importantly, mutation of this site resulted in a seven- to ten-fold increase in cyclin D1 half-life and a remarkable resistance to polyubiquitylation [26].

Subsequent studies suggested a model wherein phosphorylation of this conserved threonine served as a signal for ubiquitin-mediated degradation. However, several questions remained unanswered, including the identity of the protein kinase responsible for phosphorylating this residue and the relationship, if any, between this phosphorylation event and growth factor signaling. A key to answering these questions was the finding that cyclin D1 phosphorylation occurred in the absence of growth factor signaling (and that ubiquitin-dependent degradation of cyclin D1 can occur in both mitogen-replete and mitogen-depleted cells). This information revealed a critical property of the putative kinase. By definition, its activity should be increased in the absence of growth factors and inhibited by growth factor stimulation. Another key insight was provided through the generation of point mutations adjacent to Thr-286. Such analyses revealed a role for the adjacent proline, Pro-287, as a requisite for Thr-286 phosphorylation, suggesting that the kinase was "proline directed" and should be antagonized by mitogenic stimulation. Among the prototypical proline-directed protein kinases are CDKs themselves, as well as MAPK members. However, most if not all these protein kinase family members are induced by growth factor-derived signals. Indeed, direct analysis revealed that neither MAPK nor CDKs had this residue as their main target. Surprisingly, the relevant kinase was one primarily identified as a regulator of glycogen metabolism with, at the time, an under-appreciated role in cell growth regulation. The protein kinase, glycogen synthase kinase-3 beta (GSK3β), exhibited all the properties that were predicted: (1) it phosphorylated cyclin D1 specifically on Thr-286 in a Pro-287-dependent manner, (2) it was active in quiescent cells, and (3) its expression/activity was downregulated by growth factor stimulation [28]. Critically, the regulation of GSK3β also involves the Ras-PI3K and AKT signaling pathways (Fig. 4.1), revealing the mechanism whereby growth factors communicate directly with the cell division machinery to direct cell proliferation. The finding that GSK3β regulates cyclin D1 turnover revealed the central importance of Ras as a transducer of mitogen-mediated growth in that, in addition to regulating *CCND1* mRNA expression through MAPK signaling, it functions to maintain protein accumulation through an AKT-dependent regulation of GSK3β.

Protein kinases other than GSK3β have been reported to regulate cyclin D1. However, the role of these kinases may be limited to specific tissues or may be mediating responses to specific stimuli. Indeed, ERK1 activity is dependent upon growth factors and is not active in quiescent cells. In addition, biochemical reconstitution experiments suggested that it could not directly phosphorylate cyclin D1. In addition, the p38 MAPK family member may contribute to phosphorylation of D-type cyclins under certain conditions. For example, cyclin D3, which is also regulated by phosphorylation of an analogous threonine residue, appears to be a p38 substrate [29].

4.4 Interdependence of Cyclin D1 Polyubiquitylation, Phosphorylation, and Nucleocytoplasmic Export

Although cyclin D1 accumulates in the nucleus during the G1 phase of the cell cycle, it localizes to the cytoplasm during the remaining interphase [5, 31]. As the critical functions of cyclin D1, such as activation of CDK4 and phosphorylation of pRB, require nuclear localization, the redistribution of cyclin D1 complexes to the cytoplasm following G1 implies that regulation of the nucleocytoplasmic distribution of cyclin D1 is critical for cellular homeostasis.

Transit of proteins between the nuclear and cytoplasmic compartments occurs via nuclear pores [48, 49]. During cytoplasmic-to-nuclear translocation through the nuclear pore, cargo proteins must first interact with soluble, cytoplasmic import factors (importins), followed by the docking of this complex to the cytoplasmic face of the nuclear pore. Translocation of the importin/cargo complex through the nuclear pore requires the Ras-related GTPase Ran in its GDP bound state (Ran-GDP) on the cytoplasmic face and Ran-GTP, along with *n*uclear *t*ransport *f*actor 2 (NTF2), on the nucleoplasmic face of the nuclear pore [49]. Once the importin/cargo complex reaches the nucleoplasm, the cargo is released through binding of Ran-GTP to importin-β, resulting in the release of importin-α as well as the cargo [26, 50].

The mechanisms governing nuclear export are less well defined. However, one exportin, CRM1, which binds to leucine-rich nuclear export signals (NES), has been identified [51, 52]. Data available suggest that nuclear-to-cytoplasmic movement also depends upon the Ran-GDP/Ran-GTP gradient [48]. These findings provide a framework for understanding how the cell regulates subcellular localization of proteins. Protein movement is bidirectional (nuclear import and export), and consequently, the subcellular distribution of most proteins depends on a balance between nuclear import and nuclear export processes.

Although the precise mechanisms governing the nuclear import and export of cyclin D1 have been elusive, it has been shown that phosphorylation of Thr-286 by GSK3β promotes the nucleo-cytoplasmic redistribution of wild-type cyclin D1 [31]. GSK3β-dependent phosphorylation of cyclin D1 is required for the binding of cyclin D1 to the nuclear exportin CRM1, which is in turn necessary for the transit of cyclin D1 to the cytoplasm following completion of G1 phase (Fig. 4.3). This work provides a direct link between growth factor signaling, ubiquitin-dependent degradation, and subcellular localization.

Why is cyclin D1 relocated to the cytoplasm? Based on the observation that cyclin D1 moves to the cytoplasm precisely at the point when cyclin D1 should undergo degradation, it was suggested that ubiquitylation of the protein may in fact occur in the cytoplasm. However, evidence to support this supposition first required the identification of an E3 ligase that specifically directs the polyubiquitylation to phospho-Thr-286 of cyclin D1. Fbxo4, a member of the F-box family of substrate adaptors belonging to the Skp1-Cul1-F-box (SCF) family of E3 ligases, was subsequently identified (along with αB-crystallin) as the potential E3 ligase involved

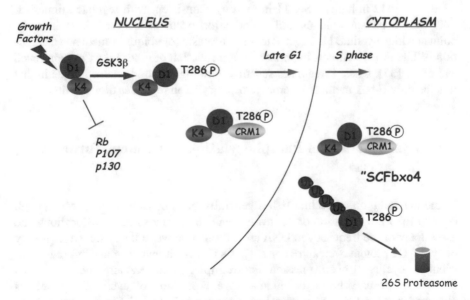

Fig. 4.3 Phosphorylation of Thr-286 directs cyclin D1 nuclear export and cytoplasmic polyubiquitin-dependent degradation

Fig. 4.4 SCF$^{Fbx4-\alpha Bcrystallin}$
E3 ligase

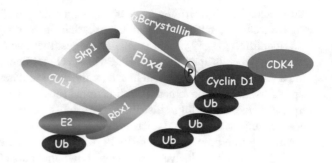

(Fig. 4.4). F-box proteins, such as Fbxo4, function as substrate adaptors that typically recognize unique degrons upon substrate phosphorylation [30]. While SCF
complexes can localize to either the nucleus or the cytoplasm, SCF^{Fbxo4} was noted
to localize exclusively to the cytoplasm, providing an explanation for the necessity
of cyclin D1 export prior to its ubiquitin-dependent degradation [31]. Thus, Thr-
286 phosphorylation of cyclin D1 provides two essential signals (Fig. 4.3), a signal
that directs cyclin D1 to the cytoplasm, where it is accessible to its E3 ligase, and
a second signal that is required for its binding to SCF^{Fbxo4}, which is mediated by
Fbxo4 and αB-crystallin.

While the regulation of cyclin D1 by direct phosphorylation has become well
established, further work suggested that the contribution of GSK3β to G1/S transition is not limited to its direct role in cyclin D1 accumulation. Thus, phosphorylation of Fbxo4 by GSK3β also regulates Fbxo4 dimerization and activity [32].
Specifically, Fbxo4 dimerization requires phosphorylation of a conserved serine
residue (Ser-12 in human, Ser-11 in mouse). Consistent with regulated dimerization, Ser-12 phosphorylation and dimerization occur at the G1/S boundary, the
point at which cyclin D1 is targeted to the cytoplasm for ubiquitin-mediated destruction. While dimerization is important for the function of other F-box proteins such
as Fbxw7 [33], where it regulates substrate binding, Fbxo4 remains unique in that
it is the only F-box member wherein phosphorylation regulates dimerization.

4.5   Dysregulation of Phosphorylation-Dependent Turnover in Cancer

While deregulation of cyclin D1 has generally been considered to occur through
either 11q13 amplification or its involvement in chromosomal translocations, the
assessment of the frequency in DNA aberrations relative to the observed frequency
of cyclin D1 protein overexpression suggests that additional mechanisms must contribute to cyclin D1 overexpression. For example, while gene amplification occurs
in 13% of primary breast carcinomas, overexpression of cyclin D1 protein is
observed in at least 50% of primary breast cancers [17, 23, 53]. Similar observations
have been made in pituitary tumors [22], hereditary nonpolyposis colorectal cancer

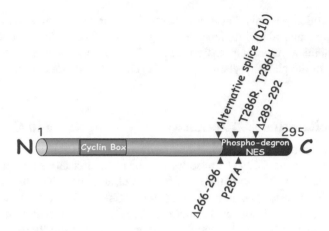

Fig. 4.5 Cancer-derived cyclin D1 mutants

Fig. 4.6 Fbx4 mutations identified in esophageal carcinoma

(HNCC) [54], and head and neck carcinomas [49]. In addition, while overexpression is often observed, experimental systems support a model wherein it is nuclear overexpression and constitutive activation of the cyclin D1/CDK4 kinase that is the key event that triggers neoplastic growth [34, 35]. If this is the case, based upon the previous discussion, it would seem likely that cyclin D1 would be often subject to mutations that disrupt nuclear export and cytoplasmic degradation.

This prediction has indeed been validated genomically through ongoing cancer genome sequencing efforts (Fig. 4.5). Mutations within cyclin D1 have been noted in numerous primary cancers and tumor-derived cell lines, including endometrial [36] and esophageal carcinomas [55]. Human cancer genome sequencing has uncovered additional mutant allele in uterine cancer, melanoma, and head and neck cancers. The majority of these mutations cluster within the C-terminus of cyclin D1, specifically to sequences that direct nuclear export and ubiquitin-dependent degradation, including Thr-286. In addition to mutations within *CCND1*, Fbxo4 is also targeted by inactivating mutations in human cancer, consistent with its function as a tumor suppressor gene. In esophageal cancers, these mutations cluster around the regulatory phosphorylation site that directs dimerization (Fig. 4.6). Fbxo4 inactivation correlates with nuclear upregulation of cyclin D1/CDK4 kinase complexes. Importantly, this implies once again that it is the dysregulation of nuclear cyclin D1, and hence cyclin D1/CDK4 activity, that is the key responsible for neoplastic growth. In addition to mutations in Thr-286 of cyclin D1, or within Fbxo4 coding region, as discussed in the following sections, *CCND1* is also subject to tumor-associated alternative splicing events that remove exon 5, which encodes residues that direct nuclear export and ubiquitin-dependent degradation. The collective data gathered from biochemical/molecular analyses from human cancers therefore reveal the importance of the phosphorylation status of cyclin D1 in neoplastic growth.

4.6 Mechanisms of Alternative *CCND1* Splicing in Cancer

The concept that the *CCND1* transcript is subject to alternative splicing was first noted in 1995 [56], following the identification of an alternative transcript, named "transcript b". Transcript b arises as a result of a failure to splice the exon 4/intron 4 boundary, giving rise to a novel protein, Cyclin D1b, which lacks exon 5-encoded residues involved in cyclin D1 regulation (e.g., T286 and the PEST domain) and harbors a novel 33-amino-acid-long C-terminal region (Fig. 4.7) [51, 56]. Thus, the prediction was that this variant would undergo altered subcellular localization as a function of a cell cycle. This hypothesis was validated by subsequent studies, which identified cyclin D1b as a constitutively nuclear protein in a number of model systems [57, 58]. Nonetheless, the overall stability of the protein was not significantly different from that of the full-length cyclin D1. Thus, the half-lives of cyclin D1 and cyclin D1b appear to be quite similar [57, 58], suggesting that even though cyclin D1b fails to be governed by Thr-286-mediated phosphorylation – which affects nuclear export and degradation – alternative and yet undefined mechanisms must exist to regulate cyclin D1 turnover.

The mechanisms that regulate alternative splicing and cyclin D1b production have garnered much recent attention. Initial findings indicated that multiple factors likely contribute to the alternative splicing event. Both full-length cyclin D1- and cyclin D1b-encoding transcripts have been detected in normal cells [52], but the data so far suggest that the alternative transcript is most prevalent in human

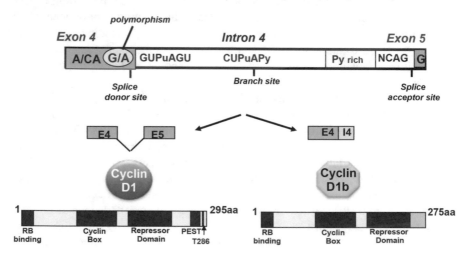

Fig. 4.7 Structure of exonic/intronic sequences implicated in D1/D1b splicing

malignancies. Interestingly, the splice donor site that regulates exon 4/intron 4 splicing encompasses a polymorphism (G/A870) that has been tentatively associated with cancer risk [56, 59–61]. In principle, the G870 allele creates an ideal splicing donor site (91% consensus sequence, using a weighted splice site identification matrix), whereas the A870 allele creates a less efficient splicing donor site (85% consensus) [48]. It was therefore hypothesized that the presence of A870 may facilitate the retention of intron 4, which would be read-through during translation to generate cyclin D1b. The potential impact of this postulate was further supported by selected genome-wide studies suggesting that the A870 allele was associated with enhanced cancer risk, as well as studies linking the A870 allele with enhanced transcript b levels [50, 52, 56]. However, subsequent studies focusing on primary human tissues to assess the impact of the G/A870 polymorphism revealed a complex role for this allele in cyclin D1b production. These studies confirmed that, indeed, the A-allele is associated with transcript b production in nonmalignant tissues. However, the impact of the A-allele is lost in tumor tissues, suggesting that the upregulation of cyclin D1b observed in human cancers must be regulated by factors independent of (or in addition to) the G/A870 allele [59].

To date, at least two alternative mechanisms have been identified that may enhance the alternative splicing event and cyclin D1b production in tumor cells, each one of these mechanisms being associated with aberrations in splicing factors. Initially, it was noted that the RNA-binding protein and splicing factor SRSF1 (formerly known as SF2, ASF, or SRp30a) bind directly to transcript b and modulates cyclin D1b production [48]. Using mini-genes of the exon 4/intron 4 boundary, it was shown that transcript b production and SRSF1 association with transcript b were reduced in the presence of the A-allele, therefore suggesting that SRSF1 binding facilitates intron 4 inclusion [48]. This concept was further supported by the observation that SRSF1 predominantly associates with, and facilitates expression of, transcript b from the G870 allele [48]. However, in primary human tumors, the influence of the A-allele was obviated as compared to that seen in non-neoplastic tissue [59]. Rather, studies in human prostate cancers revealed that SRSF1 is frequently upregulated in these transformed tissues, strongly correlating with cyclin D1b production, independently of the G/A870 allele status [48]. Thus, while SRSF1 cooperates with the G870, in tumor cells SRSF1 acts in an allele-independent manner to boost intron-4 retention and resultant cyclin D1b production. Notably, SRSF1 was recently identified as a proto-oncogene, but the underlying basis of its oncogenic activity remains poorly defined [37]. It will be important to discern how cyclin D1b production contributes to SRSF1-mediated cellular transformation.

In addition to SRSF1, the Sam68 splicing factor has also been implicated in alternative splicing of the *CCND1* transcript, with cyclin D1b production, in the context of human malignancies [38]. Sam68 is a member of the STAR (signal transduction and activation of RNA metabolism) family of RNA-binding proteins [39]. Similar to SRSF1, Sam68 is highly expressed in neoplastic tissues, and RNA immunoprecipitation studies similarly identified Sam68 in association with transcript b of *CCND1* [38]. Subsequent in vitro studies determined that Sam68 promotes cyclin D1b production, with a preference for the A870 allele [38]. However, the ability of Sam68 to favor transcript b production was enhanced by Ras or MEK/ERK activity, thus implicating pro-mitogenic and oncogenic signaling pathways as effectors of cyclin D1b production [38]. Furthermore, Sam68 was found to bind to intron 4 in a manner that is mutually exclusive with the U1–70 K spliceosomal complex, which suggest a model wherein Sam68 interferes with the recruitment of the splicing machinery, and through this mechanism, promotes cyclin D1b production [38]. These functions are not mutually exclusive with SRSF1 activity, and the collective observations indicate that both the G/A870 allele and tumor-associated alterations in expression or activity of the splicing factor cooperate to enhance expression of cyclin D1b.

4.7 Cyclin D1b: Impact on Cancer Phenotypes and Therapeutic Intervention

A litany of studies has demonstrated that cyclin D1b holds overlapping but distinct functions (when compared to full-length cyclin D1), functions that seem to enhance the pro-tumorigenic activity of cyclin D1. In multiple models engineered to over-express individual cyclin D1 isoforms, it was observed that cyclin D1b (but not full-length cyclin D1) confers an enhanced capacity for foci formation, anchorage-independent growth, and in vivo tumor formation [40, 57, 58]. These observations suggest that, in matched contexts, cyclin D1b is a gain-of-function variant of cyclin D1 with enhanced transformation activity. Interestingly, the oncogenic functions of cyclin D1b appear to be different from those associated with cell cycle regulation. Although cyclin D1b binds CDK4, early analyses suggested that cyclin D1b/CDK4 complexes have a reduced capacity to phosphorylate pRB when produced in vitro, although complexes purified from cells retain high levels of activity [57, 58]. Of interest, expression of cyclin D1b does not correlate with enhanced proliferative capacity (Ki67 indices) in human tumors [41].

Given the different transforming properties of cyclin D1 and cyclin D1b, there is currently a major effort in the field to discern the distinctive oncogenic functions of cyclin D1b. Persistent nuclear localization of the variant is likely to contribute and, in this sense, much has been learned from T286A mutations, which mimic this event [42]. Notably, nuclear cyclin D1 has been linked to increased expression and activity of the PRMT5 methyltransferase, thus inducing cyclin D1-dependent expression of the licensing factor CDT1, aberrations in DNA synthesis, and genomic instability [4, 43]. Recent studies concordantly show that ATM perturbations enhance the oncogenic capacity of nuclear cyclin D1, promote genomic instability, and select for alterations in the murine c-Myc locus in vivo [44].

In addition to these observations using the constitutive allele, parallel studies examining cyclin D1b function revealed alterations of cyclin D1 transcriptional regulation that appear to enhance tumor development and progression. Thus, consistent with the mounting body of literature linking cyclin D1 to transcriptional regulation (through both direct transcription factor binding and CDK-mediated phosphorylation of transcriptional regulators), proteomic analyses identified cyclin D1 in complex with a large number of key transcriptional regulatory molecules and sequence-specific transcription factors [45]. While cyclin D1 has been identified in association with chromatin and as a modulator of gene expression in numerous human tumor tissues, the biological impact of this transcriptional function has been specially relevant in hormone-dependent cancers [51]. Cyclin D1 is known to bind to, and modulate the function of, estrogen receptor alpha (ERα) [46, 47]. Further investigations revealed that cyclin D1b is altered in its capacity to bolster ERα activity. This is notable, as cyclin D1b expression is accompanied by resistance to ERα-directed therapies in model systems of disease [9]; in human breast cancers, cyclin D1b positivity is also associated with poor prognosis [11]. Interestingly, rectal tumorigenesis is markedly enhanced by cyclin D1b expression in female animals [12], suggesting a close interplay between hormonal status and cyclin D1b that may also contribute to progression of this tumor type.

In the context of prostate cancer, the ability of cyclin D1b to alter the function of the androgen receptor (AR) appears to be a major contributor to cyclin D1b's oncogenic capacity [13, 40]. AR activity is required during the development and progression of prostate cancers [14]. Using model systems of cyclin D1b induction, as well as morpholino-induced cyclin D1b-to-cyclin D1 conversion, it was observed that cyclin D1b "redirects" AR to different sites, in addition to alter AR activity [40]. One key event underlying disease phenotypes is the action of cyclin D1b in directing AR to the *SNAI2* (Slug) regulatory locus with the resultant AR-dependent induction of Slug expression [40]. Through this pathway, cyclin D1b contributes to signaling events that are sufficient to induce metastases in vivo and are associated with lethal disease in clinical settings [40]. In breast and bladder cancer cells, cyclin D1b has also been observed to promote migration and invasion, albeit through alternative pathways [15, 16]. Based on these and related findings, a major aspect of cyclin D1b in promoting tumor phenotypes appears to occur through altered transcriptional regulation.

The concept that cyclin D1b enhances the oncogenic properties of various hormone-dependent tumor cells, promotes resistance to hormone therapies, and facilitates metastasis has gained further support from clinical observations. Induction of cyclin D1b has been reported in a large number of tumor types, including Ewing's sarcoma [17], mantle cell lymphoma [19–21], esophageal cancer [57, 60], colon cancer [22], B-lymphoid malignancies [23], and breast cancer [9, 41]. In addition to the clinical findings linking cyclin D1b to metastases in prostate cancer, cyclin D1b has also been identified as a predictive factor for therapeutic response in colorectal cancer [22]. Based on these and other preclinical findings, the potential of cyclin D1b to serve as a biomarker to predict tumor progression and therapeutic response deserves further scrutiny.

4.8 Alternative Splicing of Cyclin D2

Is there any evidence for the existence and production of splicing-derived isoforms of cyclin D2 and D3? Several reports have emerged to suggest a role for alternative *CCND2* splicing in human disease. First reported in 2003, viral integration of the Graffi murine leukemia virus into the Gris1 locus occurs with high frequency [24]. Strikingly, integration into this site induces both overexpression and alternative splicing of the *CCND2* transcript. The resulting shorten (17 kDa) isoform of cyclin D2 has not been well characterized functionally, but their association with myeloid leukemia in mouse models is intriguing [24]. Subsequent modeling of the 17 kDa species revealed an ability to transform primary murine fibroblasts in combination with Ras. Strikingly, however, this isoform of cyclin D2 proved to be a poor catalyst of CDK-mediated pRB phosphorylation, suggesting that the mechanisms used by this short form to promote transformation may be distinct from that of canonical cyclin D2 [62]. Further insight into the oncogenic significance of the alternative splicing of the *CCND2* transcript was gained by unbiased in vitro analyses wherein mRNA isoforms with shortened 3'UTRs were detected in cancer cells [53, 62]. Shorter isoforms were shown to typically increase mRNA stability and protein production. Interestingly, the mechanisms involved include a bypass in miRNA-mediated RNA degradation and alternative polyadenylation [53]. Indeed, "shorter" forms of the *CCND2* transcript resulted in enhanced cyclin D2 production and S phase progression [53].

By contrast, a different cyclin D2 splice form, named cyclin D2SV, was isolated from murine heart tissue [54]. This cyclin D2 variant is expressed in the embryonic myocardium and appears to play a role in cell cycle exit. Unlike cyclin D1b, cyclin D2SV is retained in the endoplasmic reticulum, Golgi bodies, and lysosomes and may serve as a "sink" to sequester factors involved in cell cycle progression [49]. Consistent with this hypothesis, the negative impact of cyclin D2SV on cell cycle progression can be overcome by the overexpression of D- or B-type cyclins [54]. Overall, these observations put forward the provocative hypothesis that alternative splicing of D-type cyclins may be used as a mechanism either to promote cell cycle progression and/or tumor phenotypes (as observed with cyclin D1, the 17 kDa cyclin D2 short isoform, or those isoforms generated from transcripts with a shortened 3'UTR) or, alternatively, to attenuate cell cycle progression (as observed with cyclin D2SV).

4.9   Conclusion(s)

The initial identification of *CCND1* as a gene that is often targeted for translocation in human cancer cells, in conjunction with the demonstration that cyclin D1 expression was regulated by growth factor receptor signaling, has reinforced the role of cyclin D1 as a driver of cancer. While initial efforts focused on how growth pathways regulated *CCND1* gene expression, it has become increasingly apparent that posttranslational modifications play a major role in the regulation of the pro-tumorigenic activities of cyclin D1. The discussion in this chapter has emphasized that, besides the increased accumulation of cyclin D1, nuclear export and ubiquitin-mediated turnover of the protein also represent key regulatory events that might be altered in cancer. Importantly, both events are determined by phosphorylation on Thr-286 of cyclin D1. In the absence of phosphorylation, cyclin D1/CDK4 complexes have unrestricted access to the nucleus and nuclear substrates. The nuclear retention elicits a neomorphic biological function, which drives tumor initiation and growth.

In the context of cancer, alternative splicing of *CCND1* transcript removes the phospho-degron and thus results in a nuclear, oncogenic splice variant (cyclin D1b) which is analogous to point mutations in the cyclin D1 phospho-degron. While cyclin D1b and cyclin D1T286A have some similar qualities, such as nuclear retention, cyclin D1b is not as stable as tumor-derived point mutants, suggesting that some intronic sequences direct an alternative mechanism of degradation, perhaps also restricting its biochemical and biological function. Although Sam68 and SRSF1 both contribute to cyclin D1b expression, mechanistic insights into cancer-specific splicing require additional investigation as their elucidation holds promise as potential nodes for therapeutic intervention. Regarding the current therapies that target CDK4/6, there is no evidence to suggest that the presence of any cyclin D1 mutant will alter drug efficacy. Thus, one might expect that tumors harboring alterations in cyclin D1 will still exhibit exquisite sensitivity to small molecules, such as palbociclib (PD0332991).

References

1. Matsushime H, Roussel MF, Ashmun RA, Sherr CJ. Colony-stimulating factor 1 regulates novel cyclins during the G1 phase of the cell cycle. Cell. 1991;65(4):701–13. PubMed PMID: 1827757.
2. Xiong Y, Connolly T, Futcher B, Beach D. Human D-type cyclin. Cell. 1991;65(4):691–9. PubMed PMID: 1827756.
3. Rosenberg CL, Wong E, Petty EM, Bale AE, Tsujimoto Y, Harris NL, et al. PRAD1, a candidate BCL1 oncogene: mapping and expression in centrocytic lymphoma. Proc Natl Acad Sci U S A. 1991;88(21):9638–42. PubMed PMID: 1682919.
4. Aggarwal P, Lessie MD, Lin DI, Pontano L, Gladden AB, Nuskey B, et al. Nuclear accumulation of cyclin D1 during S phase inhibits Cul4-dependent Cdt1 proteolysis and triggers p53-dependent DNA rereplication. Genes Dev. 2007;21(22):2908–22. doi:10.1101/gad.1586007. Epub 2007/11/17. PubMed PMID: 18006686; PubMed Central PMCID: PMC2049193.
5. Aggarwal P, Vaites LP, Kim JK, Mellert H, Gurung B, Nakagawa H, et al. Nuclear cyclin D1/CDK4 kinase regulates CUL4 expression and triggers neoplastic growth via activation of the PRMT5 methyltransferase. Cancer Cell. 2010;18(4):329–40. doi:10.1016/j.ccr.2010.08.012. Epub 2010/10/19.PubMed PMID: 20951943; PubMed Central PMCID: PMC2957477.
6. Casimiro MC, Crosariol M, Loro E, Ertel A, Yu Z, Dampier W, et al. ChIP sequencing of cyclin D1 reveals a transcriptional role in chromosomal instability in mice. J Clin Invest. 2012;122(3):833–43. doi:10.1172/JCI60256. Epub 2012/02/07. PubMed PMID: 22307325; PubMed Central PMCID: PMC3287228.
7. Jirawatnotai S, Hu Y, Livingston DM, Sicinski P. Proteomic identification of a direct role for cyclin d1 in DNA damage repair. Cancer Res. 2012;72(17):4289–93. doi:10.1158/0008-5472.CAN-11-3549. PubMed PMID: 22915759; PubMed Central PMCID: PMC3432743.
8. Diehl JA. Cycling to cancer with cyclin Dl. Cancer Biol Ther. 2002;1(3):226–31. PubMed PMID: 12432268.
9. Wang Y, Dean JL, Millar EK, Tran TH, McNeil CM, Burd CJ, et al. Cyclin D1b is aberrantly regulated in response to therapeutic challenge and promotes resistance to estrogen antagonists. Cancer Res. 2008;68(14):5628–38. doi:68/14/5628 [pii] 10.1158/0008-5472.CAN-07-3170. Epub 2008/07/18. PubMed PMID: 18632615.

10. Motokura T, Bloom T, Kim HG, Juppner H, Ruderman JV, Kronenberg HM, et al. A novel cyclin encoded by a bcl1-linked candidate oncogene. Nature. 1991;350(6318):512–5. PubMed PMID: 1826542.

11. Abramson VG, Troxel AB, Feldman M, Mies C, Wang Y, Sherman L, et al. Cyclin D1b in human breast carcinoma and coexpression with cyclin D1a is associated with poor outcome. Anticancer Res. 2010;30(4):1279–85. doi:30/4/1279 [pii]. Epub 2010/06/10. PubMed PMID: 20530440; PubMed Central PMCID: PMC3874215.

12. Kim CJ, Tambe Y, Mukaisho K, Sugihara H, Isono T, Sonoda H, et al. Female-specific rectal carcinogenesis in cyclin D1b transgenic mice. Carcinogenesis. 2013;35(1):227–36. doi:bgt293 [pii] 10.1093/carcin/bgt293. Epub 2013/08/27. PubMed PMID: 23975835.

13. Burd CJ, Petre CE, Morey LM, Wang Y, Revelo MP, Haiman CA, et al. Cyclin D1b variant influences prostate cancer growth through aberrant androgen receptor regulation. Proc Natl Acad Sci U S A. 2006;103(7):2190–5. doi:0506281103 [pii] 10.1073/pnas.0506281103. Epub 2006/02/08. PubMed PMID: 16461912; PubMed Central PMCID: PMC1413684.

14. Knudsen KE, Scher HI. Starving the addiction: new opportunities for durable suppression of AR signaling in prostate cancer. Clin Cancer Res. 2009;15(15):4792–8. doi:1078–0432. CCR-08-2660 [pii] 10.1158/1078-0432.CCR-08-2660. Epub 2009/07/30. PubMed PMID: 19638458; PubMed Central PMCID: PMC2842118.

15. Kim CJ, Nishi K, Isono T, Okuyama Y, Tambe Y, Okada Y, et al. Cyclin D1b variant promotes cell invasiveness independent of binding to CDK4 in human bladder cancer cells. Mol Carcinog. 2009;48(10):953–64. doi:10.1002/mc.20547. Epub 2009/05/06. PubMed PMID: 19415719.

16. Wu FH, Luo LQ, Liu Y, Zhan QX, Luo C, Luo J, et al. Cyclin D1b splice variant promotes alphavbeta3-mediated adhesion and invasive migration of breast cancer cells. Cancer Lett. 2014;355(1):159–67. doi:S0304–3835(14)00497–2 [pii] 10.1016/j.canlet.2014.08.044. Epub 2014/09/07. PubMed PMID: 25193465.

17. Sanchez G, Bittencourt D, Laud K, Barbier J, Delattre O, Auboeuf D, et al. Alteration of cyclin D1 transcript elongation by a mutated transcription factor up-regulates the oncogenic D1b splice isoform in cancer. Proc Natl Acad Sci U S A. 2008;105(16):6004–9. doi:0710748105 [pii] 10.1073/pnas.0710748105. Epub 2008/04/17. PubMed PMID: 18413612; PubMed Central PMCID: PMC2329709.

18. Okano J, Snyder L, Rustgi AK. Genetic alterations in esophageal cancer. Methods Mol Biol. 2003;222:131–45. doi:10.1385/1-59259-328-3:131. Epub 2003/04/25.PubMed PMID: 12710684.

19. Carrere N, Belaud-Rotureau MA, Dubus P, Parrens M, de Mascarel A, Merlio JP. The relative levels of cyclin D1a and D1b alternative transcripts in mantle cell lymphoma may depend more on sample origin than on CCND1 polymorphism. Haematologica. 2005;90(6):854–6. Epub 2005/06/14. PubMed PMID: 15951302.

20. Krieger S, Gauduchon J, Roussel M, Troussard X, Sola B. Relevance of cyclin D1b expression and CCND1 polymorphism in the pathogenesis of multiple myeloma and mantle cell lymphoma. BMC Cancer. 2006;6:238. doi:1471–2407–6-238 [pii] 10.1186/1471-2407-6-238. Epub 2006/10/07. PubMed PMID: 17022831; PubMed Central PMCID: PMC1609182.

21. Marzec M, Kasprzycka M, Lai R, Gladden AB, Wlodarski P, Tomczak E, et al. Mantle cell lymphoma cells express predominantly cyclin D1a isoform and are highly sensitive to selective inhibition of CDK4 kinase activity. Blood. 2006;108(5):1744–50. doi:10.1182/blood-2006-04-016634. Epub 2006/05/13. PubMed PMID: 16690963; PubMed Central PMCID: PMC1895502.

22. Myklebust MP, Li Z, Tran TH, Rui H, Knudsen ES, Elsaleh H, et al. Expression of cyclin D1a and D1b as predictive factors for treatment response in colorectal cancer. Br J Cancer. 2012;107(10):1684–91. doi:bjc2012463 [pii] 10.1038/bjc.2012.463. Epub 2012/10/27. PubMed PMID: 23099809; PubMed Central PMCID: PMC3493874.

23. Leveque C, Marsaud V, Renoir JM, Sola B. Alternative cyclin D1 forms a and b have different biological functions in the cell cycle of B lymphocytes. Exp Cell Res. 2007;313(12):2719–29.

doi:S0014–4827(07)00198-X [pii] 10.1016/j.yexcr.2007.04.018. Epub 2007/05/15. PubMed PMID: 17499716.

24. Denicourt C, Kozak CA, Rassart E. Gris1, a new common integration site in Graffi murine leukemia virus-induced leukemias: overexpression of a truncated cyclin D2 due to alternative splicing. J Virol. 2003;77(1):37–44. Epub 2002/12/13. PubMed PMID: 12477808; PubMed Central PMCID: PMC140601.

25. Sherr CJ. Cancer cell cycles. Science. 1996;274(5293):1672–7. PMID: 8939849

26. Diehl JA, Zindy F, Sherr CJ. Inhibition of cyclin D1 phosphorylation on threonine-286 prevents its rapid degradation via the ubiquitin-proteasome pathway. Genes Dev. 1997;11(8):957–72. PubMed PMID: 9136925.

27. Wilk S, Orlowski M. Evidence that pituitary cation-sensitive neutral endopeptidase is a multicatalytic protease complex. J Neurochem. 1983;40(3):842–9. PubMed PMID: 6338156.

28. Diehl JA, Cheng M, Roussel MF, Sherr CJ. Glycogen synthase kinase-3beta regulates cyclin D1 proteolysis and subcellular localization. Genes Dev. 1998;12(22):3499–511. PubMed PMID: 9832503.

29. Ma YFQ, Sekula D, Diehl JA, Freemantle S, Dmitrovsky E. Retinoid targeting of different D-type cyclins through distinct chemopreventative mechanisms. Cancer Res. 2005;65:6467–83.

30. Skowyra D, Craig KL, Tyers M, Elledge SJ, Harper JW. F-box proteins are receptors that recruit phosphorylated substrates to the SCF ubiquitin-ligase complex. Cell. 1997;91(2):209–19. PubMed PMID: 9346238.

31. Lin DI, Barbash O, Kumar KG, Weber JD, Harper JW, Klein-Szanto AJ, et al. Phosphorylation-dependent ubiquitination of cyclin D1 by the SCF(FBX4-alphaB crystallin) complex. Mol Cell. 2006;24(3):355–66. doi:10.1016/j.molcel.2006.09.007. Epub 2006/11/04. PubMed PMID: 17081987; PubMed Central PMCID: PMC1702390.

32. Barbash O, Zamfirova P, Lin DI, Chen X, Yang K, Nakagawa H, et al. Mutations in Fbx4 inhibit dimerization of the SCF(Fbx4) ligase and contribute to cyclin D1 overexpression in human cancer. Cancer Cell. 2008;14(1):68–78. doi:S1535–6108(08)00190–6 [pii] 10.1016/j.ccr.2008.05.017. Epub 2008/07/05. PubMed PMID: 18598945; PubMed Central PMCID: PMC2597358.

33. Welcker M, Clurman BE. Fbw7/hCDC4 dimerization regulates its substrate interactions. Cell Div. 2007;2:7. PubMed PMID: 17298674.

34. Alt JR, Cleveland JL, Hannink M, Diehl JA. Phosphorylation-dependent regulation of cyclin D1 nuclear export and cyclin D1-dependent cellular transformation. Genes Dev. 2000;14(24):3102–14. PubMed PMID: 11124803.

35. Gladden AB, Diehl JA. The cyclin D1-dependent kinase associates with the pre-replication complex and modulates RB.MCM7 binding. J Biol Chem. 2003;278(11):9754–60. PubMed PMID: 12519773.

36. Moreno-Bueno G, Rodriguez-Perales S, Sanchez-Estevez C, Hardisson D, Sarrio D, Prat J, et al. Cyclin D1 gene (CCND1) mutations in endometrial cancer. Oncogene. 2003;22(38):6115–8. PubMed PMID: 12955092.

37. Karni R, de Stanchina E, Lowe SW, Sinha R, Mu D, Krainer AR. The gene encoding the splicing factor SF2/ASF is a proto-oncogene. Nat Struct Mol Biol. 2007;14(3):185–93. doi:nsmb1209 [pii] 10.1038/nsmb1209. Epub 2007/02/21. PubMed PMID: 17310252.

38. Paronetto MP, Cappellari M, Busa R, Pedrotti S, Vitali R, Comstock C, et al. Alternative splicing of the cyclin D1 proto-oncogene is regulated by the RNA-binding protein Sam68. Cancer Res. 2009;70(1):229–39. doi:0008–5472.CAN-09-2788 [pii] 10.1158/0008-5472. CAN-09-2788. Epub 2009/12/24. PubMed PMID: 20028857; PubMed Central PMCID: PMC2884274.

39. Lukong KE, Richard S. Sam68, the KH domain-containing superSTAR. Biochim Biophys Acta. 2003;1653(2):73–86. doi:S0304419X03000337 [pii]. Epub 2003/12/04. PubMed PMID: 14643926.

40. Augello MA, Burd CJ, Birbe R, McNair C, Ertel A, Magee MS, et al. Convergence of oncogenic and hormone receptor pathways promotes metastatic phenotypes. J Clin Invest. 123(1):493–508. doi:64750 [pii] 10.1172/JCI64750. Epub 2012/12/22. PubMed PMID: 23257359; PubMed Central PMCID: PMC3533295.
41. Millar EK, Dean JL, McNeil CM, O'Toole SA, Henshall SM, Tran T, et al. Cyclin D1b protein expression in breast cancer is independent of cyclin D1a and associated with poor disease outcome. Oncogene. 2009;28 (15):1812–20. doi:onc200913 [pii] 10.1038/onc.2009.13. Epub 2009/03/17. PubMed PMID: 19287456; PubMed Central PMCID: PMC3073345.
42. Kim JK, Diehl JA. Nuclear cyclin D1: an oncogenic driver in human cancer. J Cell Physiol. 2009;220(2):292–6. doi:10.1002/jcp.21791. Epub 2009/05/06. PubMed PMID: 19415697; PubMed Central PMCID: PMC2874239.
43. Aggarwal P, Vaites LP, Kim JK, Mellert H, Gurung B, Nakagawa H, et al. Nuclear cyclin D1/CDK4 kinase regulates CUL4 expression and triggers neoplastic growth via activation of the PRMT5 methyltransferase. Cancer Cell. 2010;18(4):329–40. doi:S1535–6108(10)00309–0 [pii] 10.1016/j.ccr.2010.08.012. Epub 2010/10/19. PubMed PMID: 20951943; PubMed Central PMCID: PMC2957477.
44. Vaites LP, Lian Z, Lee EK, Yin B, DeMicco A, Bassing CH, et al. ATM deficiency augments constitutively nuclear cyclin D1-driven genomic instability and lymphomagenesis. Oncogene. 2014;33(1):129–33. doi:onc2012577 [pii] 10.1038/onc.2012.577. Epub 2013/01/16. PubMed PMID: 23318439; PubMed Central PMCID: PMC4112739.
45. Bienvenu F, Jirawatnotai S, Elias JE, Meyer CA, Mizeracka K, Marson A, et al. Transcriptional role of cyclin D1 in development revealed by a genetic-proteomic screen. Nature. 2010;463(7279):374–8. doi:nature08684 [pii] 10.1038/nature08684. Epub 2010/01/22. PubMed PMID: 20090754; PubMed Central PMCID: PMC2943587.
46. Neuman E, Ladha MH, Lin N, Upton TM, Miller SJ, DiRenzo J, et al. Cyclin D1 stimulation of estrogen receptor transcriptional activity independent of cdk4. Mol Cell Biol. 1997;17(9):5338–47. PubMed PMID: 9271411.
47. Zwijsen RM, Wientjens E, Klompmaker R, van der Sman J, Bernards R, Michalides RJ. CDK-independent activation of estrogen receptor by cyclin D1. Cell. 1997;88(3):405–15. PubMed PMID: 9039267.
48. Olshavsky NA, Comstock CE, Schiewer MJ, Augello MA, Hyslop T, Sette C, et al. Identification of ASF/SF2 as a critical, allele-specific effector of the cyclin D1b oncogene. Cancer Res. 2010;70(10):3975–84. doi:0008–5472.CAN-09-3468 [pii] 10.1158/0008-5472. CAN-09-3468. Epub 2010/05/13. PubMed PMID: 20460515; PubMed Central PMCID: PMC2873684.
49. Wafa K, MacLean J, Zhang F, Pasumarthi KB. Characterization of growth suppressive functions of a splice variant of cyclin D2. PLoS One. 8(1):e53503. doi:10.1371/journal.pone.0053503 PONE-D-12-30355 [pii]. Epub 2013/01/18. PubMed PMID: 23326442; PubMed Central PMCID: PMC3542336.
50. Holley SL, Parkes G, Matthias C, Bockmuhl U, Jahnke V, Leder K, et al. Cyclin D1 polymorphism and expression in patients with squamous cell carcinoma of the head and neck. Am J Pathol. 2001;159(5):1917–24. doi:S0002–9440(10)63038–6 [pii] 10.1016/S0002-9440(10)63038-6. Epub 2001/11/07. PubMed PMID: 11696452; PubMed Central PMCID: PMC1867042.
51. Knudsen KE. The cyclin D1b splice variant: an old oncogene learns new tricks. Cell Div. 2006;1:15. doi:1747–1028–1-15 [pii] 10.1186/1747-1028-1-15. Epub 2006/07/26. PubMed PMID: 16863592; PubMed Central PMCID: PMC1559605.
52. Howe D, Lynas C. The cyclin D1 alternative transcripts [a] and [b] are expressed in normal and malignant lymphocytes and their relative levels are influenced by the polymorphism at codon 241. Haematologica. 2001;86(6):563–9. PubMed PMID: 11418364.
53. Mayr C, Bartel DP. Widespread shortening of 3'UTRs by alternative cleavage and polyadenylation activates oncogenes in cancer cells. Cell. 2009;138(4):673–84. doi:S0092–8674(09)00716–8 [pii] 10.1016/j.cell.2009.06.016. Epub 2009/08/26. PubMed PMID: 19703394; PubMed Central PMCID: PMC2819821.

54. Sun Q, Zhang F, Wafa K, Baptist T, Pasumarthi KB. A splice variant of cyclin D2 regulates cardiomyocyte cell cycle through a novel protein aggregation pathway. J Cell Sci. 2009;122(Pt 10):1563–73. doi: jcs.047738 [pii] 10.1242/jcs.047738. Epub 2009/04/30. PubMed PMID: 19401331.
55. Benzeno S, Lu F, Guo M, Barbash O, Zhang F, Herman JG, et al. Identification of mutations that disrupt phosphorylation-dependent nuclear export of cyclin D1. Oncogene. 2006;25(47):6291–303. PubMed PMID: 16732330.
56. Betticher DC, Thatcher N, Altermatt HJ, Hoban P, Ryder WD, Heighway J. Alternate splicing produces a novel cyclin D1 transcript. Oncogene. 1995;11(5):1005–11. PubMed PMID: 7675441.
57. Lu F, Gladden AB, Diehl JA. An alternatively spliced cyclin D1 isoform, cyclin D1b, is a nuclear oncogene. Cancer Res. 2003;63(21):7056–61. PubMed PMID: 14612495.
58. Solomon DA, Wang Y, Fox SR, Lambeck TC, Giesting S, Lan Z, et al. Cyclin D1 splice variants. Differential effects on localization, RB phosphorylation, and cellular transformation. J Biol Chem. 2003;278(32):30339–47. doi:10.1074/jbc.M303969200 M303969200 [pii]. Epub 2003/05/15. PubMed PMID: 12746453.
59. Comstock CE, Augello MA, Benito RP, Karch J, Tran TH, Utama FE, et al. Cyclin D1 splice variants: polymorphism, risk, and isoform-specific regulation in prostate cancer. Clin Cancer Res. 2009;15(17):5338–49. doi:1078-0432.CCR-08-2865 [pii] 10.1158/1078-0432. CCR-08-2865. Epub 2009/08/27. PubMed PMID: 19706803; PubMed Central PMCID: PMC2849314.
60. Gupta VK, Feber A, Xi L, Pennathur A, Wu M, Luketich JD, et al. Association between CCND1 G/A870 polymorphism, allele-specific amplification, cyclin D1 expression, and survival in esophageal and lung carcinoma. Clin Cancer Res. 2008;14(23):7804–12. doi:14/23/7804 [pii] 10.1158/1078-0432.CCR-08-0744. Epub 2008/12/03. PubMed PMID: 19047108; PubMed Central PMCID: PMC2723959.
61. Izzo JG, Wu TT, Wu X, Ensor J, Luthra R, Pan J, et al. Cyclin D1 guanine/adenine 870 polymorphism with altered protein expression is associated with genomic instability and aggressive clinical biology of esophageal adenocarcinoma. J Clin Oncol. 2007;25(6):698–707. doi:25/6/698 [pii] 10.1200/JCO.2006.08.0283. Epub 2007/02/20. PubMed PMID: 17308274.
62. Denicourt C, Legault P, McNabb FA, Rassart E. Human and mouse cyclin D2 splice variants: transforming activity and subcellular localization. Oncogene. 2008;27(9):1253–62. doi:1210750 [pii] 10.1038/sj.onc.1210750. Epub 2007/09/18. PubMed PMID: 17873913.

Chapter 5
Cyclin D1, Metabolism, and the Autophagy-Senescence Balance

Claudio Valenzuela and Nelson E. Brown

Abstract Progression through the cell cycle must be coordinated with crucial cell fate decisions, including the ability of a cell to exit the cell cycle and differentiate. Not surprisingly, deregulation of the G1/S transition is a well-established hallmark of cancer. While the basic mechanisms involved in this transition have been extensively characterized, it is now evident that components of the core cell cycle machinery, including cyclin D1, are functionally integrated into complex signaling and metabolic pathways not always directly related to cell cycle. In cells at risk of becoming cancerous, this complexity may underlie the cellular variability in the specific tumor suppressive processes that are implemented in response to oncogenic insults. Among these processes, autophagy has generated much debate because it may serve both as a tumor suppressive and as a pro-survival mechanism depending on the stage of tumor formation or the cell type under scrutiny. Nevertheless, a better understanding of the role of autophagy in tumorigenesis, and the functional connection of autophagy with the cell cycle and the metabolic status of the cell, may be necessary for the implementation of more rational regimens to treat cancer. In particular, recent reports have begun to unravel cyclin D1's involvement in the regulation of the autophagy-senescence balance, as well as the role of cyclin D1 function in metabolic responses. The emerging picture is concordant with the idea that cyclin D1 participates in the integration and transduction of inputs provided by both growth factors and metabolic substrates. The proper integration of these signals, in turn, may be necessary to achieve an appropriate proliferative response. To what extent these functions are exclusively dependent on cyclin D1's ability to bind and activate CDK4/CDK6, however, remains unclear.

Keywords Cell cycle • Autophagy • Metabolism • Cyclin D1 • Senescence

C. Valenzuela, PhD • N.E. Brown, MD, PhD (✉)
Center for Medical Research, University of Talca School of Medicine, Talca, Chile
e-mail: nbrown@utalca.cl

© Springer International Publishing AG 2018   111
P.W. Hinds, N.E. Brown (eds.), *D-type Cyclins and Cancer*,
Current Cancer Research, DOI 10.1007/978-3-319-64451-6_5

5.1 Introduction

Alterations in the regulatory circuits that govern the G1-S cell cycle transition are universal features of cancer [39]. At the center of these circuits are cyclin-dependent kinases (CDKs), a group of serine/threonine kinases that require the binding of short-lived *cyclins* in order to become catalytically active [19, 68, 69]. In mammalian cells, the G1 CDKs, CDK4 and CDK6, form active complexes with D-type cyclins (cyclin D1, D2, and D3) in early G1, whereas CDK2 becomes activated by E-type cyclins (cyclins E1 and E2 in mammalian cells) in late G1 [69]. Classic substrates for CDK4/CDK6- and CDK2-containing complexes are members of the so-called "pocket protein" family of repressors, which include the retinoblastoma protein (pRB), p107 and p130. Of these substrates, pRB has become the prototype and, so far, the only one directly involved in human cancer [18]. According to the most accepted model, mitogenic signals dependent on growth factors result in an increase in the expression or the half-life of D-type cyclins, allowing the formation of active cyclin D-CDK4/CDK6 complexes. These complexes, together with cyclin E-CDK2 complexes formed in late G1, help to secure the complete phosphorylation-mediated inactivation of pRB, a step necessary for the derepression or release of E2F factors responsible for driving the G1-S transition [68, 100]. While this simple model still holds true, there is evidence that, depending on the cell type or the experimental conditions used, cyclin D-CDK4/CDK6 complexes may also phosphorylate a collection of substrates not directly involved in cell cycle regulation. These substrates include transcription factors with cell-type specific functions during differentiation (e.g., FoxM1, SMAD3, members of the RUNX family, GATA-4, and MEF-2), chromatin-modifying proteins (e.g., MEP50), and proteins involved in DNA repair processes (e.g., BRCA1) [2, 7, 14, 50, 73, 90].

Given its role in G1-S transition, it is hardly surprising that cyclin D1 can be found overexpressed in a wide spectrum of human cancers, including a large proportion of luminal-type breast tumors [8, 76, 81]. In fact, recent analyses have confirmed *CCDN1*, the gene encoding cyclin D1, as one of the most frequently amplified loci in human cancer genomes [12]. Most often, however, overexpression of cyclin D1 in cancer cells takes place in the absence of any detectable genomic alteration [76]. As one would anticipate, common mechanisms of overexpression that do not involve genomic alterations include the activation of growth factor-dependent signaling pathways upstream of cyclin D1 or the loss of micro-RNAs that normally target cyclin D1 for degradation [3, 11, 15, 29, 59]. For example, most *ERBB2*-expressing human breast cancers display moderate to strong cyclin D1 expression [3, 87], and, conversely, mice lacking cyclin D1 are resistant to breast cancer induced by *ERBB2* [114]. Moreover, Choi et al. [23] showed that acute deletion of cyclin D1 blocks the progression of *ERBB2*-driven mammary tumors, an indication that the continued presence of cyclin D1 is necessary to sustain tumor growth in this model [23]. Taken together, these observations are in agreement with a model in which cyclin D1 serves as an integrator of growth-promoting signals, in such a way that its levels in early and mid-G1 dictate the probability of S-phase entrance.

It was long assumed that the oncogenic properties of cyclin D1 (and other D-type cyclins) were mostly dependent on its ability to activate CDK4 and CDK6. However, several observations and experimental findings have challenged this presumption. For example, increased levels of cyclin D1 only moderately correlate with pRB inactivation and proliferation in human tumors [1, 32, 57]. In fact, some of the oncogenic properties of D-type cyclins may depend on its ability to participate in processes other than the cell cycle in a CDK-independent manner [76, 82]. For example, cyclin D1 (like other D-type cyclins) can interact with a variety of proteins in a CDK-independent manner [25, 34], and the resulting cyclin D1-containing complexes seem to participate in cellular functions as diverse as transcriptional regulation, DNA repair, cell migration, and protein folding [14, 57, 61, 74, 79, 119]. Therefore, besides cell cycle control, deregulated expression of cyclin D1 may also affect other cellular processes in ways that could have important oncogenic consequences. It seems therefore likely that the relative contribution of the CDK-associated function of cyclin D1 to tumorigenesis may vary depending on the cell type or the specific constellation of accompanying genetic alterations (Fig. 5.1).

In spite of these complexities, the in vivo data do suggest that the ability of cyclin D1 to bind and activate CDK4/CDK6 is still relevant for tumor formation, at least in some experimental settings. For example, mammary epithelial cells derived from knockout mice that are deficient in CDK4 (the main G1 CDK that forms complexes with cyclin D1 in the mouse mammary epithelium) are resistant to *ERBB2*-dependent breast cancer in a manner that is similar to the tumor resistance reported for cyclin D1-deficient mice [115]. Overall, these reports suggested that the formation of cyclin D1-CDK4 complexes, and presumably their associated kinase activities, is required for *ERBB2*-dependent neoplastic transformation in the mammary epithelium. However, it is important to notice that cyclin D1-CDK4 complexes may also play a non-catalytic function by way of sequestering members of the WAP/CIP family of CDK inhibitors, particularly p21^{WAP1} and p27^{KIP1}, away from cyclin E-CDK2 complexes [21, 91, 92]. This means that the resistance to *ERBB2*-induced cancer observed in cyclin D1- or CDK4-deficient mammary tissues could in part be a consequence of the inhibition of cyclin E-CDK2 complexes by p27^{KIP1} or p21^{WAP1} in the absence of titrating complexes (see Fig. 5.1b). Therefore, in order to demonstrate that *ERBB2*-driven tumors are specifically dependent on the ability of cyclin D1 to activate CDK4/CDK6 (i.e., its kinase-associated function), more refined mouse models were needed.

In order to dissect the kinase-dependent functions of cyclin D1 in vivo, Landis et al. [58] generated a *knockin* mouse carrying a single amino acid substitution at the CDK4 binding region of cyclin D1 [58]. While this mutant protein (cyclin D1-K112E) still binds to CDK4 or CDK6, thus retaining the titrating function of cyclin D1-CDK4/CDK6 complexes, these complexes become enzymatically inactive [10, 42]. Importantly, similar to cyclin D1- and CDK4-deficient mice [95, 115], these *kinase dead* mice (referred to as *cyclin D1$^{KE/KE}$* mice) are also resistant to breast cancer initiated by *ERBB2* [58]. This finding demonstrates that the kinase-dependent function of cyclin D1 is necessary for tumor formation, at least breast tumor formation that is dependent on *ERBB2*. Interestingly, a further characterization

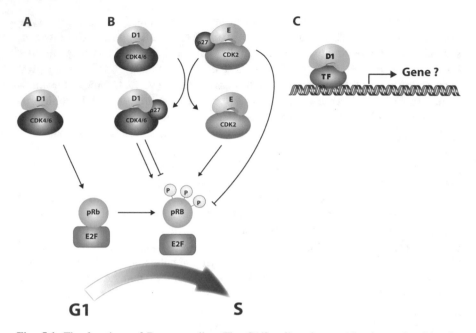

Fig. 5.1 The functions of D-type cyclins. The G1/S cell cycle transition is regulated by the sequential activation of CDK4, CDK6, and CDK2. CDK4 and CDK6 (*CDK4/6*) are activated by D-type cyclins in early G1, whereas CDK2 is activated by E-type cyclins in late G1. As shown in (**a**) and (**b**), both complexes contribute to the phosphorylation-mediated inactivation of pRB, a step necessary for entering S-phase. The activities of CDK4 or CDK6 may be inhibited by members of the INK4 family of CDK inhibitors (p16[INK4a], p15[INK4b], p18[INK4c], and p19[INK4d], not shown here), which act by competing with D-type cyclins. Likewise, members of the CIP1/KIP1 family of inhibitors (p21[WAF1], p27[KIP1], and p57[KIP2]) form inhibitory complexes with CDK2 and cyclin E. Under some circumstances, p21[WAF1] and p27[KIP1] also contribute to the stabilization and, therefore, activation of cyclin D-CDK4/6 complexes. In this case, cyclin D-CDK4/6 complexes may titrate p21[WAF1] and p27[KIP1] away from cyclin E-CDK2 complexes, thus allowing CDK2 activation (**b**). As shown in (**c**), D-type cyclins in general, and cyclin D1 in particular, can interact with a variety of proteins in a CDK-independent manner and the resulting cyclin D-containing complexes seem to participate in cellular functions as diverse as transcriptional regulation (shown here), DNA repair, cell migration and protein folding

of *cyclin D1[KE/KE]* mice revealed that the ablation of cyclin D1 activity in these animals led to a dramatic reduction in the number, as well as the differentiation capabilities, of a subset of mammary progenitors previously identified as the targets for *ERBB2*-mediated tumorigenesis [45]. Moreover, at the cellular level, *cyclin D1[KE/KE]* mammary epithelial cells display important alterations in the balance between cellular senescence and autophagy, two processes commonly disrupted in cancer cells [17]. Altogether, these studies have provided new links between the kinase activity dependent on cyclin D1 and cellular mechanisms involved in tumorigenesis, namely, self-renewal of progenitor cells, cellular senescence, and autophagy. Given the catabolic nature of autophagy, it has also become apparent that cyclin D1 activity might be functioning as a regulator of metabolism in some cell types. As explained later in this chapter, autophagy elicited in response to reduced cyclin D1-associated kinase activity could serve as a potential target for cancer treatment, although the long-term metabolic consequences of targeting autophagy are complex and highly context dependent.

5.2   The G1-S Cell Cycle Transition, Cyclin D1, and Autophagy

The ability of cancer cells to adjust their metabolism to the energy and biosynthetic demands imposed by cell proliferation is a well-known hallmark of cancer and an important contributor to anticancer drug resistance [16, 39, 107]. So far, extensive *metabolic reprograming* has been documented in connection with cellular processes commonly altered in cancer, including cellular senescence, autophagy, and stem cell self-renewal [30, 49, 104, 108]. Nevertheless, the mechanisms linking cell cycle deregulation (as observed in the context of cyclin D1 overexpression) and metabolic reprogramming, as well as the consequences of this metabolic reprogramming for the adaptation of cancer cells to their microenvironment, remain poorly characterized. For example, activation of the pRB pathway is one of the first steps in the implementation of cellular senescence [77], yet the functional links between the metabolic changes associated with cellular senescence and tumor suppression remain unclear. It is envisioned that a better understanding of these functional relationships will provide novel targets that can be used in the development of more efficacious anticancer drugs. In the following sections, we provide evidence that deregulation of the pRB pathway impinges on autophagy and metabolism. In particular, new evidence connecting cyclin D1 function, autophagy, and metabolism will be discussed.

5.2.1   Autophagy

Autophagy is a highly dynamic, evolutionarily conserved, catabolic process that involves the sequestration, and subsequent lysosomal-mediated degradation, of organelles and long-lived proteins and protein complexes [35, 53, 108]. The morphological hallmark of autophagy is the formation of double-membrane vacuoles, also known as autophagosomes, which transport cytoplasmic cargo to the lysosomes for degradation and substrate recycling [102] (Fig. 5.2). The entire process of autophagy involves the orderly assembly of more than 30 autophagy-related (ATG) gene products, each functioning in the implementation of a distinct step [41]. Among these steps, the formation of autophagosomes is probably the best characterized at the molecular level (Fig. 5.2).

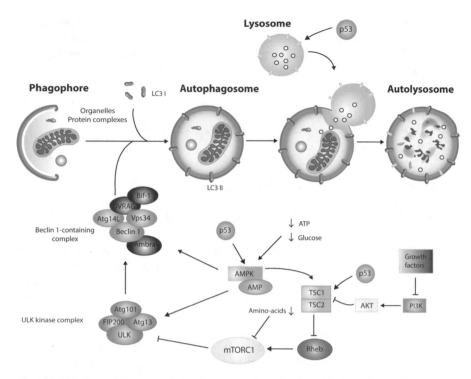

Fig. 5.2 The dynamic process of autophagy. Autophagy begins with the sequestration of organelles and long-lived proteins or protein complexes into rudimentary membranous structures known
as phagophores, which subsequently mature into LC3B-containing autophagosomes. A cytosolic
form of LC3B (*LC3-I*) is conjugated to phosphatidyl-ethanolamine to form LC3-phosphatidyl-
ethanolamine conjugate (*LC3-II*), which is recruited to autophagosomal membranes.
Autophagosomes fuse with lysosomes to form autolysosomes, and intra-autophagosomal components are degraded by lysosomal hydrolases. At the same time, LC3-II in the autolysosomal lumen
is degraded. Thus, lysosomal turnover of the autophagosomal marker LC3-II reflects starvation-
induced autophagic activity, and detecting LC3B by immunoblotting or immunofluorescence has
become a reliable method for monitoring autophagy and autophagy-related processes. The
phagophore-autophagosome transition is partially regulated by a Beclin-1 (ATG-6)/class III PI3K
(*Vps34*) complex (Beclin-1-containing complex) whose activation requires the participation of an
ULK kinase-containing complex. Autophagy is regulated by a complex signaling network, which
encompasses stimulatory and inhibitory inputs. Autophagy is also negatively regulated by the
mTORC1 (mechanistic target of rapamycin) kinase complex, a multi-protein complex that integrates both metabolic and growth-promoting signals conveyed, among others, by AMP (*AMPK*)
and PI3K/AKT kinases, respectively. Activation of growth factor receptors (such as the insulin
receptor) stimulates PI3K, leading to the activation of AKT, which inhibits the TSC1/TSC2 complex. Inhibition of this complex leads to the stabilization of the Rheb GTPase, which in turns
activates mTORC1. Activation of mTORC1 causes inhibition of autophagy through several mechanisms, including mTORC1-dependent inactivation of proteins involved in autophagosome formation (i.e., ULK1, AMBRA1, and ATG14) and the repression of transcription factors required for
lysosomal biogenesis (not shown). Energy depletion causes activation of the AMP-activated protein kinase (*AMPK*), and this event is necessary to induce autophagy in some cell types. AMPK
phosphorylates and activates TSC1/TSC2 complex, resulting in mTORC1 inhibition. AMPK also
mediates the phosphorylation-dependent activation of ULK1 and Beclin-1, two positive regulators
of autophagy. As shown here, p53 can also regulate autophagy, a function that depends on its ability to transcriptionally control various pathways that converge on mTORC1 and lysosomal regulation (not shown)

Rates of autophagy are tightly coupled to fluctuations in the intracellular concentration of specific metabolic substrates or metabolic by-products, including ATP, glucose, amino acids, fatty acids, and ammonia (one of the main by-products of amino acid catabolism) [35]. These substrates modify the activity of enzymatic complexes that function as "metabolic sensors." One of these sensors, the serine/threonine kinase mTOR (mechanistic target of rapamycin), couples nutrients and growth factor availability to cell growth and proliferation [118]. mTOR is found in two complexes, mTORC1 and mTORC2, of which mTORC1 is the most widely studied metabolic sensor [94, 118]. Other proteins or protein complexes, including AMPK (AMP-dependent protein kinase), Rag-GTPases, and Sirtuins, cooperate with, or provide inputs to, the mTOR-dependent pathway [35]. As expected, mTORC1-dependent anabolic responses are accompanied by reduced rates of autophagy. Mechanisms responsible for this reduction include mTORC1-dependent inactivation of proteins involved in autophagosome formation (i.e., ULK1, AMBRA1, and ATG14) [78, 116], as well as transcription factors (i.e., TFEB) required for lysosomal biogenesis [94]. Conversely, mTORC1 inhibition due to a multitude of starvation signals leads to higher rates of autophagy. For example, lack of glucose (which leads to reduced rates of ATP synthesis) results in the accumulation of AMP (and to a lesser extent ADP) and the subsequent activation of AMPK [40]. AMPK in turn activates TSC2, a major suppressor of mTORC1, through phosphorylation. AMPK also mediates the phosphorylation-dependent activation of ULK1 and Beclin-1, two positive regulators of autophagy [52]. Thus, reduced levels of nutrients (particularly amino acids and glucose), or the pharmacologic inhibition of mTORC1, upregulate autophagy through direct activation of factors involved in the initiation of autophagy and the induction of a lysosomal/autophagic transcriptional program [94].

Under normal nutrient conditions, basal or constitutive levels of autophagy provide a quality control mechanism that prevents the accumulation of protein aggregates or damaged organelles, thus ameliorating the endoplasmic reticulum (ER) stress response and maintaining cellular homeostasis [70–72, 108]. On the other hand, in cells subjected to starvation or other forms of stress (e.g., therapeutic stress), above-basal levels of autophagy provide basic biochemical substrates that can be utilized for energy production or to feed biosynthetic reactions, thus ensuring short-term survival [20]. Underscoring the importance of this metabolic function, autophagy-deficient mice die shortly after birth due to a failure to overcome the brief period of postnatal starvation [56]. In addition to these "pro-survival" functions, there is also evidence that under extreme conditions autophagy can serve as a mechanism of cell death [28, 31, 65]. For example, studies carried out in apoptosis-deficient mice have shown that cell lineages that would normally be eliminated through apoptosis still die while displaying an autophagic morphology, suggesting that autophagy-mediated cell death might compensate when apoptosis is compromised [93, 113].

Defects in autophagy are commonly observed in the course of aging and age-related pathologies, such as cancer and neurodegenerative diseases [22, 27]. In line with this association, global inactivation of autophagy in several animal models is accompanied by signs of premature aging, presumably as a result of the accumulation of damaged macromolecules and organelles [75].

With regard to the role of autophagy in cancer, this appears to be highly context dependent. Indeed, the available evidence suggests that autophagy may have opposite functions at different stages of tumor evolution [54, 108]. First, cells with reduced levels of autophagy, due to genetic or pharmacologic manipulations, may have a higher risk of becoming tumorigenic. This outcome is supported by several observations. For example, the hemiallelic loss of the essential autophagy gene *Beclin-1/Atg6* has been documented in up to 75% of breast, ovary, and prostate cancers [4]. Similarly, beclin-1+/− mice, which are deficient in autophagy, display an increased frequency of spontaneous malignancies [84, 117]. Moreover, Takamura et al. [97] reported the development of liver adenomas in mice carrying mosaic or liver-specific deletions of the essential autophagy regulators *ATG5* or *ATG7* [97]. Mechanistically, these effects have been linked to an impairment in the capacity of autophagy-deficient cells to degrade damaged organelles or misfolded proteins, leading to oxidative stress, tissue damage, inflammation, and genomic instability [108]. In addition to these rather indirect mechanisms of tumorigenesis, the inhibition of autophagy in some in vitro models has been shown to impair the orchestration of oncogene-induced senescence, leading directly to the acquisition of a proliferative advantage that may accelerate tumor formation [110, 112]. These observations indicate that autophagy may be involved in the orchestration of at least some of the phenotypic features of senescent cells. However, as mentioned later in this chapter, there are examples in which inhibition of autophagy correlates with the induction or exacerbation, rather than inhibition, of senescence [80], suggesting tissue- or cell-type-based variation in the response of cells to autophagy deficiency.

In contrast to the role of autophagy in suppressing tumorigenesis, other lines of evidence support the idea that autophagy may actually promote tumorigenesis by sustaining metabolism, proliferation, or survival of fully transformed cells, especially if these cells are subjected to starvation or other forms of metabolic stress [37, 109]. However, the exact metabolic consequences of inducing autophagy in these cancer cells are not well defined. Unlike quiescent or terminally differentiated cells, which are dependent on oxidative phosphorylation-mediated ATP synthesis in order to maximize energy production in conditions of limited supply of growth factors [86, 103, 107], actively proliferating cancer cells, in which growth-promoting signals are abundant, show an increase in nutrient uptake and a switch to anabolic metabolism. This metabolic reprogramming is critical to supplying nucleotides, proteins, and lipids for cell division. Moreover, this transition from oxidative phosphorylation to glycolysis, even in the presence of adequate levels of oxygen (a phenomenon known as the Warburg effect), still requires functional mitochondria for the synthesis of metabolic precursors [103, 107]. Taken together, autophagy functions as an adaptive mechanism that sustains cell viability by providing metabolic substrates for biosynthesis. It follows from this idea that inhibition of autophagy would lead to a reduction in the survival and proliferation of cancer cells. Likewise, blocking autophagy would be expected to enhance the therapeutic outcome of drugs (particularly, cancer drugs) that induce autophagy in cancer cells as a pro-survival mechanism of adaptation [6, 44].

Recently, the pro-survival role of autophagy in cancer cells has been corroborated in a variety cancer models. First, it was reported that the overexpression of oncogenic *RAS* in cancer cell lines is accompanied by high rates of autophagy. In these cells, the constant oncogenic stress associated with *RAS* activation renders mitochondrial metabolism particularly dependent on autophagy [36, 64]. Accordingly, *RAS*-expressing cancer cells in which autophagy has been blocked show a reduction in their tumor-forming capacity, which is associated with low levels of tricarboxylic acid (TCA) cycle metabolites and impaired mitochondrial function [36, 111]. Moreover, deletion of the essential autophagy genes *Atg5* or *Atg7* in a *RAS*-dependent mouse model of pancreatic cancer retards progression to high-grade intra-epithelial neoplasias and ductal adenocarcinomas in a p53-dependent manner [88]. Underscoring the role of p53 in this phenotype, deletion of p53 accelerates tumor formation in these mice. Similarly, deletion of *Atg7* in a *K-RAS*-driven mouse model of non-small-cell lung cancer (NSCLC) gives rise to more benign tumors characterized by the accumulation of defective mitochondria (oncocytomas), activation of p53, and proliferative arrest [37]. Of note, unlike the *RAS*-dependent model of pancreatic cancer, the deletion of p53 only partially rescued the tumor suppressive phenotype associated with *Atg7* loss in the lungs. From a metabolic standpoint, *Atg7*- and *p53*-deficient lung tumors display reduced fatty acid oxidation (FAO) and increased sensitivity to FAO inhibition, indicating that *RAS*-driven lung tumors require autophagy for mitochondrial function and lipid catabolism [37]. Of note, the involvement of p53 in autophagy and metabolism likely depends, at least in part, on p53's ability to modulate various pathways that converge on mTOR-containing complexes and lysosomal biogenesis [26, 33, 47, 51].

Overall, its ability to provide metabolic and biosynthetic substrates in situations of nutrient starvation, together with its ability to prevent the accumulation of damaged organelles, particularly mitochondria, renders autophagy necessary for tumorigenesis. As to the predominant cellular response to autophagy inhibition, this may vary depending on cell type and context. Such responses include apoptosis, necrosis (in cells that are deficient in apoptosis), and, most prominently, senescence. As already mentioned, given that autophagy has been considered an effector mechanism of senescence in some models [112], the fact that cellular senescence can be induced or exacerbated following autophagy inhibition is surprising and perhaps points to the existence of different types of senescence.

5.2.2   The G1/S Cell Cycle Transition and Autophagy

In order to grow and proliferate, cells must first sense and interpret a diverse collection of environmental signals. Depending on the availability and proper transduction of these signals, a decision has to be made as to whether a cell enters a reversible (quiescent) or irreversible (senescent, differentiated) form of cell cycle arrest, or simply continues to the next cycle of cell division. Intimately associated with these cell fate decisions, particularly in situations of metabolic stress, autophagy is emerging as a key process that might explain some of the adaptive consequences of cell cycle deregulation. As most of these cell fate decisions take place at the G1/S cell cycle transition, the use of mice in which regulators of this transition, including cyclin D1, were knocked out or functionally modified has been particularly informative.

An important starting point in the assessment of autophagy in mouse models of cancer was the reevaluation of some of the phenotypes displayed by *RB*-deficient mouse embryos [63]. *RB*-deficient embryos die at midgestation while exhibiting several developmental defects, including ectopic proliferation and increased apoptosis in the nervous system, lens, and liver [43]. The increased apoptosis observed in *RB*-deficient tissues was partially dependent on E2F-mediated activation of p53 [38]. Therefore, while *RB*−/−; *E2F1*−/− embryos still die in uterus, they do so at a considerable later stage of development than *RB*−/− embryos, which also correlate with a significant suppression of apoptosis, S-phase entry, and p53 activation [101]. Interestingly, at least some of the defects originally described in *RB*-deficient tissues have subsequently been attributed to hypoxia, a known inducer of autophagy, in relation to placental dysfunction [67]. In an attempt to tackle the role of pRB in hypoxic tissues, Tracy et al. reported that *RB*-deficient liver cells display signs of autophagic cell death in response to experimental hypoxia, and this effect was dependent on E2F-mediated derepression of *BNIP3*, a gene that codes for a hypoxia-inducible factor [99]. The authors of this work extended this observation to *RB*-deficient mouse embryonic fibroblasts (MEFs) and *RB*-deficient human cell lines under hypoxic conditions [99]. Similarly, isolated *RB*-deficient muscle progenitor cells (myoblasts) can still form myotubes and partially differentiate into muscle fibers in vitro but rapidly degenerate afterward, exhibiting signs of autophagy-mediated cell death [24]. Consistent with these findings, it was shown that *E2F1* overexpression directly regulates the induction of autophagy genes and enhances the rates of basal autophagy in vitro [83]. Taken together, these observations are in agreement with a model in which derepression of E2F factors secondary to *RB* loss can, in some lineages and under specific developmental circumstances such as hypoxia, tilt the balance toward the induction of autophagy rather than apoptosis as a mechanism of cell death. It must be emphasized, however, that it is presently unclear whether autophagy in these cases actually represents a failed mechanism of survival or an apoptosis-independent mechanism of cell death.

In contrast to autophagy associated to *RB* loss, Jiang et al. have shown that reintroducing *RB* into *RB*-deficient cancer cell lines also induces autophagy. In this setting, pRB binding to E2F1 (which maintains transcriptional repression) is required for autophagy induction. Accordingly, overexpression of *E2F1* overcomes this effect and tilts the balance toward the induction of apoptosis [46]. Mimicking the reintroduction of *RB*, autophagy induction was also observed following the overexpression of the cyclin-dependent kinase inhibitors (CKIs) p16^{INK4a} or p27^{KIP1}, suggesting that activation of the pRB pathway is sufficient to induce autophagy in these experimental settings [46]. This is in agreement with previous work linking overexpression of CKIs, metabolic stress, and autophagy induction [55, 62]. Thus, under conditions of metabolic stress, the phosphorylation-mediated stabilization of p27^{KIP1} by AMPK leads to autophagy upregulation. Conversely, downregulation of p27^{KIP1} under these conditions results in cell death by apoptosis, suggesting that autophagy represents a pro-survival adaptation to metabolic stress. Importantly, these effects were dependent on p27^{KIP1}-mediated modulation of CDK activity [62]. It is worth mentioning that experimental manipulations involving the restoration of *RB* or the overexpression of CKIs in *RB*-deficient or *RB*-expressing cells, respectively, are well-established models of cellular senescence [5]. Therefore, at least in some models, autophagy and senescence may indeed be part of the same tumor suppressor pathway, a possibility that was first suggested by Young et al. [112]. In this scenario, autophagy may be crucial for the implementation of complex cellular traits in senescent cells, including the senescence-associated secretory phenotype (SASP) [80]. If this is the case, autophagy-mediated catabolism might play a key part in the metabolic reprograming observed in senescent cells. Indeed, recent work has revealed a major shift to a predominantly mitochondrial, oxidative, metabolism in senescent cells [30, 49, 80]. For example, induction of senescence in human diploid fibroblasts (HDFs) following the expression of the oncogene *BRAFV600E* is associated with activation of pyruvate dehydrogenase (PDH), an enzyme that catalyzes the pyruvate-to-acetyl-CoA conversion that fuels the TCA cycle and oxidative phosphorylation [49]. Increased mitochondrial activity, oxygen consumption, ATP production, and lipid catabolism have also been documented in models of therapy-induced senescence [30]. Similarly, *RAS*-induced senescence in human fibroblasts is associated with reduced lipid synthesis, increased fatty acid oxidation, and increased oxygen consumption [85]. Taken together, these studies suggest a metabolic shift toward maximal energy production at the expense of biosynthesis in senescent cells. Whether or not autophagy contributes to this metabolic profile in all forms of senescence, however, remains a matter of debate. As discussed elsewhere, in some settings senescence can actually be induced or exacerbated upon autophagy inhibition.

5.2.3 Cyclin D1 and the Autophagy/Senescence Balance

In order to explore the cellular consequences of reducing cyclin D1 activity in the mammary epithelium, Brown et al. took advantage of the kinase dead *cyclin D1$^{KE/KE}$* mouse model. Contrary to the authors' expectations, *cyclin D1$^{KE/KE}$* mammary tissues displayed high levels of proliferation along with a failure to induce markers of senescence in response to *ERBB2* [17]. These findings indicate that aberrant proliferation can still take place in mutant tissues in response to *ERBB2* despite the presence of a canonically "active" pRB pathway, perhaps reflecting the ability of these cells to activate compensatory survival processes in order to cope with reduced levels of cyclin D1-associated kinase activity [17]. Indeed, this aberrant proliferative response was also accompanied by an upregulation of markers of autophagy. That the upregulation of autophagy in *cyclin D1$^{KE/KE}$* mammary epithelium represented a survival adaptation to reduced cyclin D1 activity was suggested by experiments carried out in an immortalized *cyclin D1*$^{KE/KE}$ cell line that retained high rates of autophagy in vitro. Thus, reducing the rates of autophagy through shRNA-mediated knockdown of *ATG5* in these cells led to an impairment of proliferation due to the reactivation of senescence [17]. Therefore, contrary to previous reports suggesting that senescence and autophagy are part of the same pathway [112], these results indicate that senescence can be induced or exacerbated by autophagy inhibition, at least in cells with reduced cyclin D1 function. Of note, induction of senescence upon autophagy inhibition has also been reported in human fibroblasts [48, 106, 112]. From a metabolic standpoint, these fibroblasts display an increased number of mitochondria and lysosomes, produce higher levels of ROS, and display a reduction in the mitochondrial membrane potential and cellular ATP content [48, 106]. In spite of these findings, the specific metabolic profiles that accompany cellular senescence in the context of autophagy inhibition will likely vary depending on the cell type and experimental context.

The link between cyclin D1 activity and the autophagy-senescence balance may also suggest a more general connection between cyclin D1 and metabolism. Thus, contrary to the view that cyclin D1 exclusively acts downstream of growth factor-derived signals to promote proliferation, functional cyclin D1-CDK4/CDK6 complexes may act as a nexus to indicate both growth factor and nutrient proficiencies appropriate for a proliferative response. In the absence of active cyclin D1-CDK4/CDK6 complexes, induction of autophagy may represent an attempt to respond to growth-promoting signals in the *perceived* absence of metabolic substrates. This inability of dysfunctional cyclin D1-CDK4/CDK6 complexes to properly sense the environment would in turn trigger a metabolic reprograming characterized by an increase in the rates of autophagy. This general model has been supported by recent reports linking cyclin D1 and metabolism in hepatocytes and mammary epithelial cells (see below). This model also suggests a close functional cooperation between cyclin D1 function and bona fide metabolic sensors and effectors, particularly mTOR-containing complexes.

5.2.4   Cyclin D1 and Metabolism

The link between cyclin D1 function and metabolic reprograming has been recently confirmed in several models [60, 82]. In hepatocytes, cyclin D1-CDK4 complexes modulate metabolic responses, independent of cell division, through the phosphorylation-mediated activation of the histone acetyltransferase GCN5 (general control non-repressed protein 5) [60]. Among other substrates, GCN5 acetylates the transcriptional coactivator PGC-1α (peroxisome-proliferator-activated receptor gamma coactivator 1 alpha), suppressing its transcriptional activity. Conversely, Sirtuin-1 deacetylates and therefore activates PGC-1α [96]. As a transcriptional coactivator, PGC-1α promotes the expression of several genes involved in gluconeogenesis and mitochondrial respiration and, at the same time, induces

ROS-detoxifying enzymes [96]. Therefore, inhibition of CDK4 or downregulation of cyclin D1 in hepatocytes increases the pool of deacetylated, active PGC-1α and leads to "fasting-like" state characterized by an increase in glucose production and utilization through transcriptional derepression of PGC-1α-dependent gluconeogenic genes [13, 60]. Conversely, insulin-mediated signaling following refeeding facilitates the formation of cyclin D1-CDK4 complexes, leading to suppression of hepatic gluconeogenesis [60] (see Fig. 5.3). Beyond the liver, morphological and functional changes indicative of metabolic reprograming have also been observed in cyclin D1−/− (null) embryonic fibroblasts (MEFs) and mammary epithelial cells. Overall, these cells display an increase in mitochondrial size and activity, with signs of reduced cytosolic glycolysis [89, 105]. Mechanistically, nuclear respiratory factor 1 (NRF-1), a transcription factor that induces nuclear-encoded mitochondrial genes, might be inactivated by cyclin D1 in a CDK-dependent manner [105]. Thus, reduced expression of cyclin D1 or CDK4, as well as blocking the activity of cyclin D1-CDK4 complexes, has the effect of increasing mitochondrial respiration at the expense of cytosolic glycolysis (Fig. 5.3). Conversely, mammary tumor cells that overexpress cyclin D1 show an inhibition of mitochondrial activity with an enhancement of cytosolic glycolysis [89]. More recently, a direct involvement of cyclin D1 in mitochondrial function has also been suggested. Thus, cyclin D1 can physically interact, in a CDK-independent manner, with the voltage-dependent anion channel (VDAC) localized at the outer mitochondrial membrane, inhibiting the transport of ATP, ADP, and other metabolites and thus impairing mitochondrial function [98].

In summary, there is compelling evidence from different experimental systems that cyclin D1-CDK complexes are involved in the integration and transduction of metabolic signals (Fig. 5.3). However, how these processes are coordinated with autophagy remains unclear.

5.3 Concluding Remarks and Future Directions

In the preceding sections, we have tried to integrate recent lines of evidence connecting cyclin D/CDK function, metabolism, and autophagy. It has become evident that the mechanisms in which autophagy is activated, as well as the specific cellular effects that autophagy activation may have, can vary depending on cell type or the specific stimulus involved. In tumors, this variability likely reflects both the nature of the mutational events that a cancer cell has already experienced and the changes of the coevolving microenvironment. Although the ultimate mechanisms by which cell cycle deregulation may affect autophagy are far from being completely understood, we speculate that part of the answer will come from a careful reevaluation of already existing animal models. This analysis will give us invaluable information about the interplay between autophagy, differentiation, senescence, and apoptosis during development in the absence of cell cycle regulators.

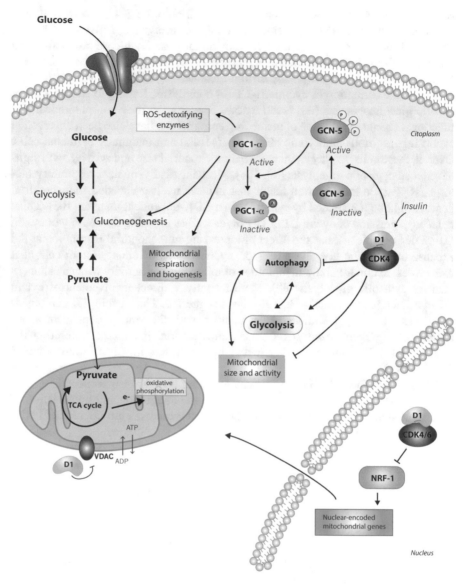

Fig. 5.3 Cyclin D1 and metabolism. Cyclin D1-CDK4 complexes modulate metabolism at least in part through phosphorylation-mediated activation of the acetyltransferase GCN5. One of the targets of GCN5 is PGC-1α, a transcriptional coactivator of several gluconeogenic genes. PGC-1α also promotes mitochondrial respiration, inducing, at the same time, ROS-detoxifying enzymes. As shown here, upon acetylation, the transcriptional function of PGC-1α is inhibited. Therefore, inhibition of CDK4 or downregulation of cyclin D1 in hepatocytes leads to a derepression of gluconeogenic genes and an increase in glucose production and utilization. On the other hand, mammary epithelial cells (MECs) specifically deficient in cyclin D1-associated kinase activity display high levels of autophagy, implying that cyclin D1-CDK4 complexes inhibit autophagy under normal conditions. There are also reports showing that cyclin D1-deficient (cyclin D1 null) mouse

As mentioned in this chapter, cyclin D1-CDK-pRB-E2F deregulation can induce or repress autophagy depending on the cell type and specific stress conditions. In particular, the contrasting outcomes observed between "primary" cells and cancer cells propagated in vitro may well be a reflection of the mutational histories of different cell lineages, a fact that needs to be considered when interpreting findings. Loss or gain of function mutations specifically designed to target members of the CDK-pRB-E2F pathway will be necessary to clarify the role of these proteins in autophagy regulation and tumorigenesis. It is plausible that many phenotypes that have been described in cell cycle mutant mice (including embryonic lethality) might need a reinterpretation in the context of regulation of autophagy and metabolism.

The last decade has witnessed important advances toward the development of specific CDK4/CDK6 inhibitors for cancer treatment [9]. As must be evident from the preceding sections, however, disrupting the cyclin D-CDK4/CDK6 complexes can trigger an extensive metabolic reprogramming, which may result, in some cell types, in an upregulation of autophagy. Taking into account these new findings, and given the dual function of autophagy during cancer initiation and progression, the incorporation of pharmacological modulators of autophagy as anticancer drugs must be cautious. Many of the current anticancer therapies, including drugs that inhibit CDK4/CDK6 kinases, have been shown to induce autophagy in tumor cells. However, there is an ongoing debate as to whether autophagy is required for the efficient killing of tumor cells following chemo- or radiotherapies or whether autophagy represents an adaptive response that enables tumor cells to survive the therapeutic insult [20]. Obviously, inhibition of autophagy will lead to opposite therapeutic outcomes depending on which one of these possibilities applies. Nonetheless, most studies seem to indicate that autophagy inhibition sensitizes tumor cells to a wide spectrum of therapies [20, 66]. Thus, a better understanding of the metabolic and growth suppressive pathways that may be enhanced by autophagy inhibition will be necessary to expand the therapeutic window of current therapeutic regimens and to confront the almost certain development of drug resistance.

Fig. 5.3 (continued) embryonic fibroblasts (MEFs) and MECs display an increase in mitochondrial size and activity, with signs of reduced cytosolic glycolysis. Mechanistically, nuclear respiratory factor 1 (*NRF-1*), a transcription factor that induces nuclear-encoded mitochondrial genes, can be inactivated by cyclin D1 in a CDK-dependent manner. Thus, reduced expression of cyclin D1 or CDK4, as well as blocking the activity of cyclin D1-CDK4 complexes, has the effect of increasing mitochondrial respiration at the expense of cytosolic glycolysis. Conversely, mammary tumor cells that overexpress cyclin D1 show an inhibition of mitochondrial activity with an enhancement of cytosolic glycolysis. Furthermore, cyclin D1 can physically interact, in a CDK-independent manner, with the voltage-dependent anion channel (*VDAC*) localized at the outer mitochondrial membrane, leading to an impairing in mitochondrial function. At present, the relationship between, on the one hand, autophagy, glycolysis, and mitochondrial activity and, on the other hand, the functional status of cyclin D1-CDK complexes is poorly understood (depicted here as *bidirectional curved arrows*)

References

1. Agarwal R, Gonzalez-Angulo AM, Myhre S, Carey M, Lee JS, Overgaard J, Alsner J, Stemke-Hale K, Lluch A, Neve RM, et al. Integrative analysis of cyclin protein levels identifies cyclin b1 as a classifier and predictor of outcomes in breast cancer. Clin Cancer Res Off J Am Assoc Cancer Res. 2009;15:3654–62.
2. Aggarwal P, Vaites LP, Kim JK, Mellert H, Gurung B, Nakagawa H, Herlyn M, Hua X, Rustgi AK, McMahon SB, et al. Nuclear cyclin D1/CDK4 kinase regulates CUL4 expression and triggers neoplastic growth via activation of the PRMT5 methyltransferase. Cancer Cell. 2010;18:329–40.
3. Ahnstrom M, Nordenskjold B, Rutqvist LE, Skoog L, Stal O. Role of cyclin D1 in ErbB2-positive breast cancer and tamoxifen resistance. Breast Cancer Res Treat. 2005;91: 145–51.
4. Aita VM, Liang XH, Murty VV, Pincus DL, Yu W, Cayanis E, Kalachikov S, Gilliam TC, Levine B. Cloning and genomic organization of beclin 1, a candidate tumor suppressor gene on chromosome 17q21. Genomics. 1999;59:59–65.
5. Alexander K, Hinds PW. Requirement for p27(KIP1) in retinoblastoma protein-mediated senescence. Mol Cell Biol. 2001;21:3616–31.
6. Amaravadi RK, Yu D, Lum JJ, Bui T, Christophorou MA, Evan GI, Thomas-Tikhonenko A, Thompson CB. Autophagy inhibition enhances therapy-induced apoptosis in a Myc-induced model of lymphoma. J Clin Invest. 2007;117:326–36.
7. Anders L, Ke N, Hydbring P, Choi YJ, Widlund HR, Chick JM, Zhai H, Vidal M, Gygi SP, Braun P, et al. A systematic screen for CDK4/6 substrates links FOXM1 phosphorylation to senescence suppression in cancer cells. Cancer Cell. 2011;20:620–34.
8. Arnold A, Papanikolaou A. Cyclin D1 in breast cancer pathogenesis. J Clin Oncol. 2005;23:4215–24.
9. Asghar U, Witkiewicz AK, Turner NC, Knudsen ES. The history and future of targeting cyclin-dependent kinases in cancer therapy. Nat Rev Drug Discov. 2015;14:130–46.
10. Baker GL, Landis MW, Hinds PW. Multiple functions of D-type cyclins can antagonize pRb-mediated suppression of proliferation. Cell Cycle. 2005;4:330–8.
11. Bandi N, Zbinden S, Gugger M, Arnold M, Kocher V, Hasan L, Kappeler A, Brunner T, Vassella E. miR-15a and miR-16 are implicated in cell cycle regulation in a Rb-dependent manner and are frequently deleted or down-regulated in non-small cell lung cancer. Cancer Res. 2009;69:5553–9.
12. Beroukhim R, Mermel CH, Porter D, Wei G, Raychaudhuri S, Donovan J, Barretina J, Boehm JS, Dobson J, Urashima M, et al. The landscape of somatic copy-number alteration across human cancers. Nature. 2010;463:899–905.
13. Bhalla K, Liu WJ, Thompson K, Anders L, Devarakonda S, Dewi R, Buckley S, Hwang BJ, Polster B, Dorsey SG, et al. Cyclin D1 represses gluconeogenesis via inhibition of the transcriptional coactivator PGC1alpha. Diabetes. 2014;63:3266–78.
14. Bienvenu F, Jirawatnotai S, Elias JE, Meyer CA, Mizeracka K, Marson A, Frampton GM, Cole MF, Odom DT, Odajima J, et al. Transcriptional role of cyclin D1 in development revealed by a genetic-proteomic screen. Nature. 2010;463:374–8.
15. Bonci D, Coppola V, Musumeci M, Addario A, Giuffrida R, Memeo L, D'Urso L, Pagliuca A, Biffoni M, Labbaye C, et al. The miR-15a-miR-16-1 cluster controls prostate cancer by targeting multiple oncogenic activities. Nat Med. 2008;14:1271–7.
16. Boroughs LK, DeBerardinis RJ. Metabolic pathways promoting cancer cell survival and growth. Nat Cell Biol. 2015;17:351–9.
17. Brown NE, Jeselsohn R, Bihani T, Hu MG, Foltopoulou P, Kuperwasser C, Hinds PW. Cyclin D1 activity regulates autophagy and senescence in the mammary epithelium. Cancer Res. 2012;72:6477–89.
18. Burkhart DL, Sage J. Cellular mechanisms of tumour suppression by the retinoblastoma gene. Nat Rev Cancer. 2008;8:671–82.

19. Casimiro MC, Crosariol M, Loro E, Li Z, Pestell RG. Cyclins and cell cycle control in cancer and disease. Genes Cancer. 2012;3:649–57.
20. Chen N, Debnath J. Autophagy and tumorigenesis. FEBS Lett. 2010;584:1427–35.
21. Cheng M, Sexl V, Sherr CJ, Roussel MF. Assembly of cyclin D-dependent kinase and titration of p27Kip1 regulated by mitogen-activated protein kinase kinase (MEK1). Proc Natl Acad Sci U S A. 1998;95:1091–6.
22. Choi AM, Ryter SW, Levine B. Autophagy in human health and disease. N Engl J Med. 2013;368:651–62.
23. Choi YJ, Li X, Hydbring P, Sanda T, Stefano J, Christie AL, Signoretti S, Look AT, Kung AL, von Boehmer H, et al. The requirement for cyclin D function in tumor maintenance. Cancer Cell. 2012;22:438–51.
24. Ciavarra G, Zacksenhaus E. Rescue of myogenic defects in Rb-deficient cells by inhibition of autophagy or by hypoxia-induced glycolytic shift. J Cell Biol. 2010;191:291–301.
25. Coqueret O. Linking cyclins to transcriptional control. Gene. 2002;299:35–55.
26. Crighton D, Wilkinson S, O'Prey J, Syed N, Smith P, Harrison PR, Gasco M, Garrone O, Crook T, Ryan KM. DRAM, a p53-induced modulator of autophagy, is critical for apoptosis. Cell. 2006;126:121–34.
27. Cuervo AM, Dice JF. Age-related decline in chaperone-mediated autophagy. J Biol Chem. 2000;275:31505–13.
28. Debnath J, Brugge JS. Modelling glandular epithelial cancers in three-dimensional cultures. Nat Rev Cancer. 2005;5:675–88.
29. Desai KV, Xiao N, Wang W, Gangi L, Greene J, Powell JI, Dickson R, Furth P, Hunter K, Kucherlapati R, et al. Initiating oncogenic event determines gene-expression patterns of human breast cancer models. Proc Natl Acad Sci U S A. 2002;99:6967–72.
30. Dorr JR, Yu Y, Milanovic M, Beuster G, Zasada C, Dabritz JH, Lisec J, Lenze D, Gerhardt A, Schleicher K, et al. Synthetic lethal metabolic targeting of cellular senescence in cancer therapy. Nature. 2013;501:421–5.
31. Elgendy M, Sheridan C, Brumatti G, Martin SJ. Oncogenic Ras-induced expression of Noxa and Beclin-1 promotes autophagic cell death and limits clonogenic survival. Mol Cell. 2011;42:23–35.
32. Ertel A, Dean JL, Rui H, Liu C, Witkiewicz AK, Knudsen KE, Knudsen ES. RB-pathway disruption in breast cancer: differential association with disease subtypes, disease-specific prognosis and therapeutic response. Cell Cycle. 2010;9:4153–63.
33. Feng Z, Zhang H, Levine AJ, Jin S. The coordinate regulation of the p53 and mTOR pathways in cells. Proc Natl Acad Sci U S A. 2005;102:8204–9.
34. Fu M, Wang C, Li Z, Sakamaki T, Pestell RG. Minireview: Cyclin D1: normal and abnormal functions. Endocrinology. 2004;145:5439–47.
35. Galluzzi L, Pietrocola F, Levine B, Kroemer G. Metabolic control of autophagy. Cell. 2014;159:1263–76.
36. Guo JY, Chen HY, Mathew R, Fan J, Strohecker AM, Karsli-Uzunbas G, Kamphorst JJ, Chen G, Lemmons JM, Karantza V, et al. Activated Ras requires autophagy to maintain oxidative metabolism and tumorigenesis. Genes Dev. 2011;25:460–70.
37. Guo JY, Karsli-Uzunbas G, Mathew R, Aisner SC, Kamphorst JJ, Strohecker AM, Chen G, Price S, Lu W, Teng X, et al. Autophagy suppresses progression of K-ras-induced lung tumors to oncocytomas and maintains lipid homeostasis. Genes Dev. 2013;27:1447–61.
38. Hakem R, Mak TW. Animal models of tumor-suppressor genes. Annu Rev Genet. 2001;35:209–41.
39. Hanahan D, Weinberg RA. Hallmarks of cancer: the next generation. Cell. 2011;144:646–74.
40. Hardie DG, Ross FA, Hawley SA. AMPK: a nutrient and energy sensor that maintains energy homeostasis. Nat Rev Mol Cell Biol. 2012;13:251–62.
41. He C, Klionsky DJ. Regulation mechanisms and signaling pathways of autophagy. Annu Rev Genet. 2009;43:67–93.
42. Hinds PW, Mittnacht S, Dulic V, Arnold A, Reed SI, Weinberg RA. Regulation of retinoblastoma protein functions by ectopic expression of human cyclins. Cell. 1992;70:993–1006.

43. Jacks T, Fazeli A, Schmitt EM, Bronson RT, Goodell MA, Weinberg RA. Effects of an Rb mutation in the mouse. Nature. 1992;359:295–300.
44. Janku F, McConkey DJ, Hong DS, Kurzrock R. Autophagy as a target for anticancer therapy. Nat Rev Clin Oncol. 2011;8:528–39.
45. Jeselsohn R, Brown NE, Arendt L, Klebba I, Hu MG, Kuperwasser C, Hinds PW. Cyclin D1 kinase activity is required for the self-renewal of mammary stem and progenitor cells that are targets of MMTV-ErbB2 tumorigenesis. Cancer Cell. 2010;17:65–76.
46. Jiang H, Martin V, Gomez-Manzano C, Johnson DG, Alonso M, White E, Xu J, McDonnell TJ, Shinojima N, Fueyo J. The RB-E2F1 pathway regulates autophagy. Cancer Res. 2010;70:7882–93.
47. Jones RG, Thompson CB. Tumor suppressors and cell metabolism: a recipe for cancer growth. Genes Dev. 2009;23:537–48.
48. Kang HT, Lee KB, Kim SY, Choi HR, Park SC. Autophagy impairment induces premature senescence in primary human fibroblasts. PLoS One. 2011;6:e23367.
49. Kaplon J, Zheng L, Meissl K, Chaneton B, Selivanov VA, Mackay G, van der Burg SH, Verdegaal EM, Cascante M, Shlomi T, et al. A key role for mitochondrial gatekeeper pyruvate dehydrogenase in oncogene-induced senescence. Nature. 2013;498:109–12.
50. Kehn K, Berro R, Alhaj A, Bottazzi ME, Yeh WI, Klase Z, Van Duyne R, Fu S, Kashanchi F. Functional consequences of cyclin D1/BRCA1 interaction in breast cancer cells. Oncogene. 2007;26:5060–9.
51. Kenzelmann Broz D, Spano Mello S, Bieging KT, Jiang D, Dusek RL, Brady CA, Sidow A, Attardi LD. Global genomic profiling reveals an extensive p53-regulated autophagy program contributing to key p53 responses. Genes Dev. 2013;27:1016–31.
52. Kim J, Guan KL. AMPK connects energy stress to PIK3C3/VPS34 regulation. Autophagy. 2013;9:1110–1.
53. Kim KH, Lee MS. Autophagy – a key player in cellular and body metabolism. Nat Rev Endocrinol. 2014;10:322–37.
54. Kimmelman AC. The dynamic nature of autophagy in cancer. Genes Dev. 2011;25:1999–2010.
55. Komata T, Kanzawa T, Takeuchi H, Germano IM, Schreiber M, Kondo Y, Kondo S. Antitumour effect of cyclin-dependent kinase inhibitors (p16(INK4A), p18(INK4C), p19(INK4D), p21(WAF1/CIP1) and p27(KIP1)) on malignant glioma cells. Br J Cancer. 2003;88:1277–80.
56. Kuma A, Hatano M, Matsui M, Yamamoto A, Nakaya H, Yoshimori T, Ohsumi Y, Tokuhisa T, Mizushima N. The role of autophagy during the early neonatal starvation period. Nature. 2004;432:1032–6.
57. Lamb J, Ramaswamy S, Ford HL, Contreras B, Martinez RV, Kittrell FS, Zahnow CA, Patterson N, Golub TR, Ewen ME. A mechanism of cyclin D1 action encoded in the patterns of gene expression in human cancer. Cell. 2003;114:323–34.
58. Landis MW, Pawlyk BS, Li T, Sicinski P, Hinds PW. Cyclin D1-dependent kinase activity in murine development and mammary tumorigenesis. Cancer Cell. 2006;9:13–22.
59. Lee RJ, Albanese C, Fu M, D'Amico M, Lin B, Watanabe G, Haines GK 3rd, Siegel PM, Hung MC, Yarden Y, et al. Cyclin D1 is required for transformation by activated Neu and is induced through an E2F-dependent signaling pathway. Mol Cell Biol. 2000;20:672–83.
60. Lee Y, Dominy JE, Choi YJ, Jurczak M, Tolliday N, Camporez JP, Chim H, Lim JH, Ruan HB, Yang X, et al. Cyclin D1-Cdk4 controls glucose metabolism independently of cell cycle progression. Nature. 2014;510:547–51.
61. Li Z, Wang C, Jiao X, Lu Y, Fu M, Quong AA, Dye C, Yang J, Dai M, Ju X, et al. Cyclin D1 regulates cellular migration through the inhibition of thrombospondin 1 and ROCK signaling. Mol Cell Biol. 2006;26:4240–56.
62. Liang J, Shao SH, Xu ZX, Hennessy B, Ding Z, Larrea M, Kondo S, Dumont DJ, Gutterman JU, Walker CL, et al. The energy sensing LKB1-AMPK pathway regulates p27(kip1) phosphorylation mediating the decision to enter autophagy or apoptosis. Nat Cell Biol. 2007;9:218–24.
63. Lipinski MM, Jacks T. The retinoblastoma gene family in differentiation and development. Oncogene. 1999;18:7873–82.

64. Lock R, Roy S, Kenific CM, Su JS, Salas E, Ronen SM, Debnath J. Autophagy facilitates glycolysis during Ras-mediated oncogenic transformation. Mol Biol Cell. 2011;22:165–78.
65. Lum JJ, Bauer DE, Kong M, Harris MH, Li C, Lindsten T, Thompson CB. Growth factor regulation of autophagy and cell survival in the absence of apoptosis. Cell. 2005;120:237–48.
66. Maclean KH, Dorsey FC, Cleveland JL, Kastan MB. Targeting lysosomal degradation induces p53-dependent cell death and prevents cancer in mouse models of lymphomagenesis. J Clin Invest. 2008;118:79–88.
67. MacPherson D, Sage J, Crowley D, Trumpp A, Bronson RT, Jacks T. Conditional mutation of Rb causes cell cycle defects without apoptosis in the central nervous system. Mol Cell Biol. 2003;23:1044–53.
68. Malumbres M, Barbacid M. To cycle or not to cycle: a critical decision in cancer. Nat Rev Cancer. 2001;1:222–31.
69. Malumbres M, Barbacid M. Cell cycle, CDKs and cancer: a changing paradigm. Nat Rev Cancer. 2009;9:153–66.
70. Mathew R, Karantza-Wadsworth V, White E. Role of autophagy in cancer. Nat Rev Cancer. 2007a;7:961–7.
71. Mathew R, Karp CM, Beaudoin B, Vuong N, Chen G, Chen HY, Bray K, Reddy A, Bhanot G, Gelinas C, et al. Autophagy suppresses tumorigenesis through elimination of p62. Cell. 2009;137:1062–75.
72. Mathew R, Kongara S, Beaudoin B, Karp CM, Bray K, Degenhardt K, Chen G, Jin S, White E. Autophagy suppresses tumor progression by limiting chromosomal instability. Genes Dev. 2007b;21:1367–81.
73. Matsuura I, Denissova NG, Wang G, He D, Long J, Liu F. Cyclin-dependent kinases regulate the antiproliferative function of Smads. Nature. 2004;430:226–31.
74. McMahon C, Suthiphongchai T, DiRenzo J, Ewen ME. P/CAF associates with cyclin D1 and potentiates its activation of the estrogen receptor. Proc Natl Acad Sci U S A. 1999;96:5382–7.
75. Mizushima N, Levine B. Autophagy in mammalian development and differentiation. Nat Cell Biol. 2010;12:823–30.
76. Musgrove EA, Caldon CE, Barraclough J, Stone A, Sutherland RL. Cyclin D as a therapeutic target in cancer. Nat Rev Cancer. 2011;11:558–72.
77. Narita M, Nunez S, Heard E, Narita M, Lin AW, Hearn SA, Spector DL, Hannon GJ, Lowe SW. Rb-mediated heterochromatin formation and silencing of E2F target genes during cellular senescence. Cell. 2003;113:703–16.
78. Nazio F, Strappazzon F, Antonioli M, Bielli P, Cianfanelli V, Bordi M, Gretzmeier C, Dengjel J, Piacentini M, Fimia GM, et al. mTOR inhibits autophagy by controlling ULK1 ubiquitylation, self-association and function through AMBRA1 and TRAF6. Nat Cell Biol. 2013;15:406–16.
79. Neuman E, Ladha MH, Lin N, Upton TM, Miller SJ, DiRenzo J, Pestell RG, Hinds PW, Dowdy SF, Brown M, et al. Cyclin D1 stimulation of estrogen receptor transcriptional activity independent of cdk4. Mol Cell Biol. 1997;17:5338–47.
80. Perez-Mancera PA, Young AR, Narita M. Inside and out: the activities of senescence in cancer. Nat Rev Cancer. 2014;14:547–58.
81. Perou CM, Sorlie T, Eisen MB, van de Rijn M, Jeffrey SS, Rees CA, Pollack JR, Ross DT, Johnsen H, Akslen LA, et al. Molecular portraits of human breast tumours. Nature. 2000;406:747–52.
82. Pestell RG. New roles of cyclin D1. Am J Pathol. 2013;183:3–9.
83. Polager S, Ofir M, Ginsberg D. E2F1 regulates autophagy and the transcription of autophagy genes. Oncogene. 2008;27:4860–4.
84. Qu X, Yu J, Bhagat G, Furuya N, Hibshoosh H, Troxel A, Rosen J, Eskelinen EL, Mizushima N, Ohsumi Y, et al. Promotion of tumorigenesis by heterozygous disruption of the beclin 1 autophagy gene. J Clin Invest. 2003;112:1809–20.
85. Quijano C, Cao L, Fergusson MM, Romero H, Liu J, Gutkind S, Rovira II, Mohney RP, Karoly ED, Finkel T. Oncogene-induced senescence results in marked metabolic and bioenergetic alterations. Cell Cycle. 2012;11:1383–92.

86. Rathmell JC, Vander Heiden MG, Harris MH, Frauwirth KA, Thompson CB. In the absence of extrinsic signals, nutrient utilization by lymphocytes is insufficient to maintain either cell size or viability. Mol Cell. 2000;6:683–92.
87. Reis-Filho JS, Savage K, Lambros MB, James M, Steele D, Jones RL, Dowsett M. Cyclin D1 protein overexpression and CCND1 amplification in breast carcinomas: an immunohisto-chemical and chromogenic in situ hybridisation analysis. Modern Pathol Off J US Can Acad Pathol. 2006;19:999–1009.
88. Rosenfeldt MT, O'Prey J, Morton JP, Nixon C, MacKay G, Mrowinska A, Au A, Rai TS, Zheng L, Ridgway R, et al. p53 status determines the role of autophagy in pancreatic tumour development. Nature. 2013;504:296–300.
89. Sakamaki T, Casimiro MC, Ju X, Quong AA, Katiyar S, Liu M, Jiao X, Li A, Zhang X, Lu Y, et al. Cyclin D1 determines mitochondrial function in vivo. Mol Cell Biol. 2006;26:5449–69.
90. Shen R, Wang X, Drissi H, Liu F, O'Keefe RJ, Chen D. Cyclin D1-cdk4 induce runx2 ubiq-uitination and degradation. J Biol Chem. 2006;281:16347–53.
91. Sherr CJ, Roberts JM. CDK inhibitors: positive and negative regulators of G1-phase progres-sion. Genes Dev. 1999;13:1501–12.
92. Sherr CJ, Roberts JM. Living with or without cyclins and cyclin-dependent kinases. Genes Dev. 2004;18:2699–711.
93. Shimizu S, Kanaseki T, Mizushima N, Mizuta T, Arakawa-Kobayashi S, Thompson CB, Tsujimoto Y. Role of Bcl-2 family proteins in a non-apoptotic programmed cell death depen-dent on autophagy genes. Nat Cell Biol. 2004;6:1221–8.
94. Shimobayashi M, Hall MN. Making new contacts: the mTOR network in metabolism and signalling crosstalk. Nat Rev Mol Cell Biol. 2014;15:155–62.
95. Sicinski P, Donaher JL, Parker SB, Li T, Fazeli A, Gardner H, Haslam SZ, Bronson RT, Elledge SJ, Weinberg RA. Cyclin D1 provides a link between development and oncogenesis in the retina and breast. Cell. 1995;82:621–30.
96. Spiegelman BM. Transcriptional control of mitochondrial energy metabolism through the PGC1 coactivators. Novartis Found Symp. 2007;287:60–63; discussion 63–69.
97. Takamura A, Komatsu M, Hara T, Sakamoto A, Kishi C, Waguri S, Eishi Y, Hino O, Tanaka K, Mizushima N. Autophagy-deficient mice develop multiple liver tumors. Genes Dev. 2011;25:795–800.
98. Tchakarska G, Roussel M, Troussard X, Sola B. Cyclin D1 inhibits mitochondrial activity in B cells. Cancer Res. 2011;71:1690–9.
99. Tracy K, Dibling BC, Spike BT, Knabb JR, Schumacker P, Macleod KF. BNIP3 is an RB/E2F target gene required for hypoxia-induced autophagy. Mol Cell Biol. 2007;27:6229–42.
100. Trimarchi JM, Lees JA. Sibling rivalry in the E2F family. Nat Rev Mol Cell Biol. 2002;3:11–20.
101. Tsai KY, Hu Y, Macleod KF, Crowley D, Yamasaki L, Jacks T. Mutation of E2f-1 suppresses apoptosis and inappropriate S phase entry and extends survival of Rb-deficient mouse embryos. Mol Cell. 1998;2:293–304.
102. Tsuchihara K, Fujii S, Esumi H. Autophagy and cancer: dynamism of the metabolism of tumor cells and tissues. Cancer Lett. 2009;278:130–8.
103. Vander Heiden MG, Cantley LC, Thompson CB. Understanding the Warburg effect: the met-abolic requirements of cell proliferation. Science. 2009;324:1029–33.
104. Viale A, Pettazzoni P, Lyssiotis CA, Ying H, Sanchez N, Marchesini M, Carugo A, Green T, Seth S, Giuliani V, et al. Oncogene ablation-resistant pancreatic cancer cells depend on mitochondrial function. Nature. 2014;514:628–32.
105. Wang C, Li Z, Lu Y, Du R, Katiyar S, Yang J, Fu M, Leader JE, Quong A, Novikoff PM, et al. Cyclin D1 repression of nuclear respiratory factor 1 integrates nuclear DNA synthesis and mitochondrial function. Proc Natl Acad Sci U S A. 2006;103:11567–72.
106. Wang Y, Wang XD, Lapi E, Sullivan A, Jia W, He YW, Ratnayaka I, Zhong S, Goldin RD, Goemans CG, et al. Autophagic activity dictates the cellular response to oncogenic RAS. Proc Natl Acad Sci U S A. 2012;109:13325–30.

107. Ward PS, Thompson CB. Metabolic reprogramming: a cancer hallmark even warburg did not anticipate. Cancer Cell. 2012;21:297–308.
108. White E. Deconvoluting the context-dependent role for autophagy in cancer. Nat Rev Cancer. 2012;12:401–10.
109. White E. Exploiting the bad eating habits of Ras-driven cancers. Genes Dev. 2013;27:2065–71.
110. White E, Lowe SW. Eating to exit: autophagy-enabled senescence revealed. Genes Dev. 2009;23:784–7.
111. Yang S, Wang X, Contino G, Liesa M, Sahin E, Ying H, Bause A, Li Y, Stommel JM, Dell'antonio G, et al. Pancreatic cancers require autophagy for tumor growth. Genes Dev. 2011;25:717–29.
112. Young AR, Narita M, Ferreira M, Kirschner K, Sadaie M, Darot JF, Tavare S, Arakawa S, Shimizu S, Watt FM, et al. Autophagy mediates the mitotic senescence transition. Genes Dev. 2009;23:798–803.
113. Yu L, Alva A, Su H, Dutt P, Freundt E, Welsh S, Baehrecke EH, Lenardo MJ. Regulation of an ATG7-beclin 1 program of autophagic cell death by caspase-8. Science. 2004;304:1500–2.
114. Yu Q, Geng Y, Sicinski P. Specific protection against breast cancers by cyclin D1 ablation. Nature. 2001;411:1017–21.
115. Yu Q, Sicinska E, Geng Y, Ahnstrom M, Zagozdzon A, Kong Y, Gardner H, Kiyokawa H, Harris LN, Stal O, et al. Requirement for CDK4 kinase function in breast cancer. Cancer Cell. 2006;9:23–32.
116. Yuan HX, Russell RC, Guan KL. Regulation of PIK3C3/VPS34 complexes by MTOR in nutrient stress-induced autophagy. Autophagy. 2013;9:1983–95.
117. Yue Z, Jin S, Yang C, Levine AJ, Heintz N. Beclin 1, an autophagy gene essential for early embryonic development, is a haploinsufficient tumor suppressor. Proc Natl Acad Sci U S A. 2003;100:15077–82.
118. Zoncu R, Efeyan A, Sabatini DM. mTOR: from growth signal integration to cancer, diabetes and ageing. Nat Rev Mol Cell Biol. 2011;12:21–35.
119. Zwijsen RM, Wientjens E, Klompmaker R, van der Sman J, Bernards R, Michalides RJ. CDK-independent activation of estrogen receptor by cyclin D1. Cell. 1997;88:405–15.

Chapter 6
Death of a Dogma: Cyclin D Activates Rb by Mono-phosphorylation

Steven F. Dowdy

Abstract The current textbook dogma of G_1 cell cycle progression proposes that cyclin D-Cdk4/Cdk6 inactivates the pRb tumor suppressor during early G_1 phase by progressive multi-phosphorylation, termed hypo-phosphorylation, to release E2F transcription factors that specifically turn on the cyclin E gene, which then activate Cdk2 and complete pRb's inactivation by hyper-phosphorylation at the restriction point. However, this model has remained largely untested from a biochemical stand-point for more than 20 years. Moreover, the biologically active form(s) of pRb present during early G_1 phase is uncharacterized, and a precise understanding of a potential "pRb phospho-code," regulating association with individual pRb partners, remains elusive. Recently, using quantitative 2D isoelectric focusing (2D IEF) to directly count phosphates on pRb, we have shown pRb to be exclusively mono-phosphorylated in early G_1 phase by cyclin D-Cdk4/Cdk6 complexes acting to modify only one of each of the 14 Cdk phosphorylation sites per individual pRb molecule. Mono-phosphorylated pRb is functionally active in early G1 phase and binds to and represses E2Fs as well as the E1a oncoprotein. Biologically, cells undergoing a DNA damage response activate cyclin D-Cdk4/Cdk6 to generate mono-phosphorylated pRb to regulate global transcription, whereas un-phosphorylated pRb is inactive during a DNA damage response. At the late G_1 restriction point, activation of cyclin E-Cdk2 complexes inactivates pRb by quantum hyper-phosphorylation. These observations fundamentally change our understanding of G_1 cell cycle progression and show that mono-phosphorylated pRb, generated by cyclin D-Cdk4/Cdk6, is the biologically active pRb isoform in early G_1 phase that represses E2F transcription and that cyclin E-Cdk2 is the pRb inactivating kinase. These findings raise clear questions about the role of cyclin D1 as an oncogene and about the cellular state elicited by drugs targeting cyclin D1's partner

S.F. Dowdy, PhD (✉)
Department of Cellular and Molecular Medicine, UCSD School of Medicine,
La Jolla, CA, USA, 92093
e-mail: sdowdy@ucsd.edu

© Springer International Publishing AG 2018   133
P.W. Hinds, N.E. Brown (eds.), *D-type Cyclins and Cancer*,
Current Cancer Research, DOI 10.1007/978-3-319-64451-6_6

kinases, Cdk4 and Cdk6. Herein, we summarize the findings leading to this paradigm shift and offer a hypothesis reconciling the apparently contradictory roles of cyclin D-Cdk4/Cdk6 as pRb activator and as a common oncogenic driver.

Keywords Rb tumor suppressor • Mono-phosphorylated pRb • Un-phosphorylated pRb • Hyper-phosphorylated pRb • Cyclin D-Cdk4/Cdk6 • Cyclin E-Cdk2 • E2F • Restriction point • Early G_1 phase

6.1 Introduction

The retinoblastoma tumor suppressor protein (pRb) functions to regulate multiple critical cellular activities, including the late G_1 checkpoint or restriction point, the DNA damage response checkpoint, cell cycle exit, and differentiation [5, 15, 18, 33]. pRb contains no detectable enzymatic activity, but instead acts as a scaffold protein that binds numerous cellular transcription factors, chromatin-remodeling proteins, and other factors [29]. The prototypical cellular pRb-binding proteins are members of the E2F transcription factor family, namely, E2Fs 1–4. During early G_1 phase, pRb binds to and represses E2F transcription factor target genes required for S-phase entrance and DNA replication [5, 15]. Several groups have determined a partial structure of pRb that shows a pseudo-dimer of dimers of a highly structured A/B box (the so-called pocket) and an N-terminal A'/B' box (Fig. 6.1) [4, 22]. pRb binds E2F transcription factors at the interface between the A/B pocket, whereas the high-avidity binding viral oncoproteins, E1a, TAg, and E7, block pRb's interaction with E2Fs by binding to the LxCxE-binding domain on the B box [24].

During cell cycle progression, pRb is regulated by phosphorylation at the hands of multiple cyclin-Cdk complexes [17]. pRb contains 14 Cdk consensus S/T-P (Ser/Thr-Pro) phosphorylation sites that are spread out on the loops between the structured A', B', A, and B domains and on the long unstructured C-terminal tail near the B box domain (Fig. 6.1). Two additional potential Cdk sites at T5 and S567 are not phosphorylated in vivo [31]. For the last 20+ years, pRb was thought to exist in three general isoforms: (1) un-phosphorylated pRb in G_0; (2) hypo-phosphorylated pRb in early G_1 phase, also referred to as "under" phosphorylated pRb or "partially" phosphorylated pRb; and (3) inactive, hyper-phosphorylated pRb, present in late G_1, S, G_2, and M phases that is readily identifiable as slower migrating species by SDS-PAGE [5, 6, 15, 23, 26, 33]. Inactive hyper-phosphorylated pRb does not bind E2Fs or viral oncoproteins. Surprisingly, given the scientific scrutiny of pRb over the last 30 years, the biochemical identification of the biologically active isoform(s) of pRb required for early G_1 phase regulation, DNA damage checkpoint control, cell cycle exit, and differentiation remained, at best, ill defined or unknown.

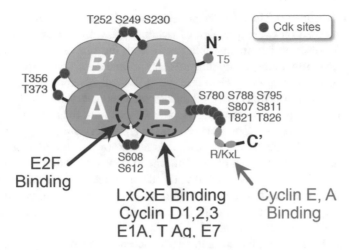

Fig. 6.1 Schematic diagram of human pRb structure showing Cdk phosphorylation sites (*red dots*), A'/B' and A/B pocket domains, E2F-binding location, cyclin D, E, and A binding locations. Note that all of the Cdk phosphorylation sites are present on unstructured loops between the four defined structural domains. The high-avidity cyclin E/A binding sites, R/KxL motifs, are present on the floppy C-terminus potentially allowing for cyclin E/A-Cdk2 to remain bound while phosphorylating all 14 Cdk sites. In contrast, the cyclin D low avidity binding site is present in the highly structured and bulky B box

6.2 Revisiting the Dogma: Cyclin D-Cdk4/Cdk6 Functionally Inactivates pRb by Progressive Hypo-phosphorylation

In most cancers outside of retinoblastoma, small cell lung carcinoma, osteosarcoma, breast cancer, and a handful of other malignancies, the *RB1* gene is infrequently mutated or deleted [5, 15]. In fact, the majority of tumors express wild-type pRb. However, upstream pathways that regulate pRb by cyclin-Cdk-mediated phosphorylation are altered in these wild-type pRb-expressing tumors, including deletion or inactivating mutations of the p16^{INK4a} tumor suppressor and increased expression or mutation of cyclin D1, D2, D3, Cdk4, and Cdk6 proteins. Starting in the mid-1990s, these observations led to the dogma of a linear "p16-cyclin D-pRb" pathway that serves to *functionally inactivate* pRb during early G_1 phase of the cell cycle, and indeed this is today's textbook view of an oncogenic pathway relevant to nearly all tumor types (Fig. 6.2) [5, 7, 15, 21, 35–37].

To dissect pRb function and regulation, many early reports utilized supraphysiologic overexpression of various cyclins (A, B, D, E) and Cdks (1, 2, 4, 6) that resulted in pRb inactivation by hyper-phosphorylation associated with an accelerated S-phase entry and induction of E2F-dependent target genes [9, 16, 27, 34]. Likewise, supraphysiologic overexpression studies using pRb constructs where many, but not all, of the putative Cdk S/T-P consensus sites were mutated to Ala residues resulted in repressed E2F-dependent transcription and cell cycle arrest, as did supraphysiologic overexpression of Cdk inhibitors, p16, p21, and p27 [5, 7, 15,

19, 21, 25, 33, 35–37]. However, because these early studies from the 1990s and early 2000s relied on overexpression of cyclins, Cdks, inhibitors, and pRb, they potentially obfuscated important subtleties in both the function and regulation of phosphorylated pRb. Indeed, in cycling primary normal cells, pRb actively represses E2F target genes during early G_1 phase when the cell is expressing p16[INK4a] and p27[Kip1], yet the cell overrides these inhibitors each and every cell cycle to activate cyclin E-Cdk2 complexes at the late G1 phase checkpoint (or restriction point) to inactivate pRb by hyper-phosphorylation and to phosphorylate p27[Kip1] for degradation. Thus, while the identification of what is now known as the pRb pathway and the functional insight into each of its components constitute a dramatic advance in cancer biology, it now seems likely that the dependence of these early studies on ectopic (over)expression of each component may have obscured an appreciation of the actual, physiologic regulation of pRb acting in both normal and tumor cells to regulate G_1 cell cycle progression.

Based on these studies, the current, widely accepted model of G_1 cell cycle progression proposes that cyclin D-Cdk4/Cdk6 functionally inactivates pRb during early G_1 phase by progressive multi-phosphorylation, called hypo-phosphorylation, resulting in release of E2F transcription factors. Such "free" E2F may then induce expression of cyclin E, resulting in activation of cyclin E-Cdk2 complexes that complete pRb inactivation by hyper-phosphorylation in late G_1 phase (Fig. 6.2). The key precept of this model is the functional inactivation of pRb by a progressive increase in the number of phosphorylated pRb residues at the hands of cyclin D-Cdk4/Cdk6

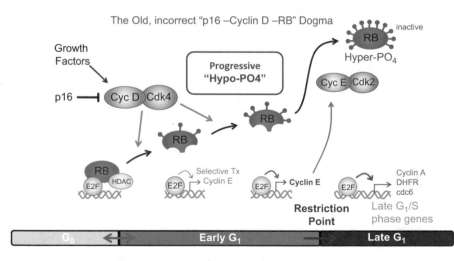

Fig. 6.2 The *current view* of G_1 cell cycle progression as it relates to the pRb pathway proposes that growth factor stimulation and activation of cyclin D-Cdk4/Cdk6 complexes results in a progressive multi-phosphorylation, termed hypo-phosphorylation, on pRb during early G_1 phase that releases some E2F transcription factors that selectively induce cyclin E, which then activates Cdk2 to "fully" inactivate pRb by hyper-phosphorylation and release of the remainder of bound E2Fs. A significant issue with this model is the lack of definitive biochemical evidence for the existence of individual pRb molecules bearing more than 1 but less than 14 modified Cdk sites

complexes. However, the putative hypo-phosphorylated pRb and un-phosphorylated pRb co-migrate on 1D SDS-PAGE and cannot be separated [11]. Moreover, there was no biochemical data defining the extent or timing of phosphorylation that constitutes hypo-phosphorylated pRb. Consequently, until late 2014, it remained entirely unknown if "hypo-phosphorylated" pRb contained one, two, three, five, seven, or more phosphates. Furthermore, it was never shown or determined how many phosphorylated residues are required to functionally inactive pRb, resulting in release of E2F transcription factors. Thus, while the fundamental concepts behind D cyclin and pRb's roles in cell cycle control in normal and tumor cells remain of general interest [7, 37], a precise biochemical accounting of the details of this model is absent, and this is in turn crucial to a proper understanding of the oncogenic actions of cyclin D1-Cdk4/Cdk6, as well as to a full appreciation of the consequences of pharmacological inhibition of these enzymes at the hands of drugs that have recently entered the clinic.

6.2.1 Death of a Dogma Part I: E2F Genes Are Not Induced in Early G_1 Phase by Cyclin D-Cdk4/Cdk6-Mediated Phosphorylation of pRb

The current view of Cdk-mediated pRb inactivation posits that cyclin D-Cdk4/Cdk6 complexes functionally inactivate pRb by progressive hypo-phosphorylation throughout early G_1 phase resulting in release of E2F transcription factors sufficient to drive expression of cyclin E, which ultimately activates Cdk2 to result in "full" phosphorylation of pRb. However, this aspect of the pRb pathway originated in the early 1990s, prior to the invention of quantitative TaqMan PCR and other quantitative RT-PCR techniques. In addition, many groups were using poor cell cycle synchronization methodologies that often resulted in a broad peak of cells at various stages of G_1 and S phase instead of highly synchronized cells. Consequently, pRb pathway models were originally built on data sets that were not as reliable as those that can now be produced by modern quantitative biochemical techniques and cell synchronization techniques.

Based on experiments using highly synchronized cells that express cyclin D, p16[INK4a], pRb, etc., at physiologic levels, by the mid-1990s, studies were showing data that began to bring into question the biological validity of the dogmatic view of the pRb pathway. Performing kinetic analyses from highly synchronized normal cells and p16[INK4a]-deficient cancer cells, including G_0 serum-deprived and serum-restimulated cells, early G_1 contact-inhibited cells in serum, or centrifugally elutriated cycling cells isolated into discrete cell cycle populations, in a series of papers we observed active cyclin D-Cdk4/Cdk6 throughout early G_1 phase simultaneous with pRb-mediated repression of E2F target genes [10–13, 30, 31, 38] (Fig. 6.3a). In fact, the only time induction of E2F target genes was observed in G_1 phase was when cyclin E-Cdk2 complexes were active and pRb was inactivated by hyper-phosphorylation.

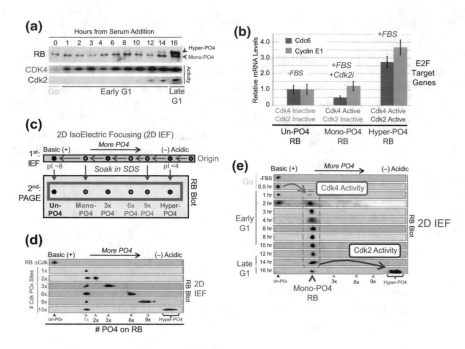

Fig. 6.3 Biochemical analysis of pRb and E2F target genes. (**a**) Kinetic analysis of primary human fibroblasts (HFFs) G_0 arrested by serum deprivation (−FBS) for 72 h, followed by serum addition (+FBS) and entrance into the cell cycle. Cells were then followed over 16 h by 1D SDS-PAGE, anit-pRb immunoblot, and anti-Cdk4/Cdk6 and anti-Cdk2 immunoprecipitation-kinase assays. (**b**) Human RPE-1 cells were serum starved for 72 h, followed by serum addition +/− Cdk2i (15 μM roscovitine), and assayed for E2F target gene induction, Cdc6 and cyclin E1, by qRT-PCR normalized to hGAPDH. Error bars: SD of three biologic replicates. (**c**) Schematic diagram of two-dimensional isoelectric focusing (2D IEF). Immunoprecipitated pRb is loaded at origin on acidic end of IEF strip and separated first by pI. IEF strip is then soaked in SDS, run in second dimension into SDS-PAGE, and immunoblotted for pRb. (**d**) 2D IEF pRb-HA immunoblot of pRb construct standards expressed in cycling cells and containing 0 (ΔCdk), 1×, 2×, 3×, 6×, 9×, or 15× Cdk phosphorylation sites. (**e**) 2D IEF pRb immunoblot from serum-deprived G_0 arrested (−FBS) and released (+FBS) primary HFFs from 0 to 16 h. Samples were same as in panel (**a**). Cyclin D-Cdk4/Cdk6 performs the mono-phosphorylation of pRb and cyclin E-Cdk2 performs the inactivating hyper-phosphorylation

To directly test whether cyclin D-Cdk4/Cdk6 and/or cyclin E-Cdk2 inactivates pRb to release E2F transcription factors, we synchronized cells by serum deprivation for 72 h, followed by serum stimulation. As observed and reported by many groups over the last 20+ years, cells in serum-free media repress E2F target genes, such as Cdc6, cyclin E1, and DHFR (Fig. 6.3b). Addition of serum activates cyclin D-Cdk4/Cdk6 complexes, followed by activation of cyclin E-Cdk2 complexes 12–16 h later culminating in the inactivation of pRb by hyper-phosphorylation and release of E2F transcription factors that induce E2F target genes to drive cells into S phase. However, this observation does not by itself reveal if Cdk4/Cdk6, Cdk2, or the combination is required for pRb inactivation for E2F binding. To address this question, after serum addition in G_1 phase, we added a Cdk2 inhibitor (Cdk2i, roscovitine) that leaves Cdk4/Cdk6 activity intact [31]. We found that pRb was fastest migrating (consistent with hypo-phosphorylation of pRb, given the active state of Cdk4/Cdk6) on 1D SDS-PAGE and E2F target genes (Cdc6 and cyclin E1) were repressed (Fig. 6.3b), demonstrating that Cdk2 is required for pRb inactivation for binding to E2Fs. Thus, in the presence of serum-stimulated, active cyclin D-Cdk4/Cdk6, pRb remains functionally active to repress E2F target genes. In other words, cyclin D-Cdk4/Cdk6 does not inactivate pRb for E2F binding and repression. Importantly, these observations brought into question a core tenet of the accepted view of the pRb pathway: putative functional inactivation of pRb by progressive hypo-phosphorylation by cyclin D-Cdk4/Cdk6 during early G_1 phase.

6.2.2 Death of a Dogma Part II: Cyclin D-Cdk4/Cdk6 Exclusively Mono-phosphorylates pRb – There Is No Such Thing as Progressive, Hypo-phosphorylated pRb

Surprisingly, from as far back as the early 1990s, the critical and fundamental missing piece of data from the "cyclin D inactivating pRb" dogma was the actual biochemical identification and quantification data of progressively hypo-phosphorylated pRb from cells (not from test-tube kinase reactions which are fraught with non-physiologic conditions). The putative hypo-phosphorylated pRb does not show any altered migration on 1D SDS-PAGE, whereas inactive hyper-phosphorylated pRb shows the classic slower migrating profile. The notion of hypo-phosphorylated pRb originated with a 1989 paper from the Livingston lab [26]. Ludlow et al. used ^{32}P orthophosphate to label endogenous pRb from cycling cells and unexpectedly found that the fastest migrating species of pRb on 1D SDS-PAGE contained phosphate, which he termed "under-phosphorylated" pRb. The extent (numbers) of phosphate groups on under-phosphorylated pRb, which eventually became known as hypo-phosphorylated pRb, remained unknown. Subsequent ^{32}P orthophosphate labeling of highly synchronized cells, combined with 2D phospho-peptide mapping analysis (where pRb is partially digested by trypsin into small peptides and separated by thin-layer chromatography) by Sibylle Mittnacht's group [28], showed that both hypo-phosphorylated pRb and hyper-phosphorylated pRb contained most of the same phosphorylation sites, albeit with a ~ 10x lower amount of phosphate on hypo-phosphorylated pRb. Unfortunately, the Mittnacht data combined with the Ludlow data were misinterpreted as proof of progressive hypo-phosphorylation of pRb on many Cdk sites, neglecting the possibility that phospho-peptide analyses of this sort cannot distinguish between a collection of singly phosphorylated pRb molecules vs multiply phosphorylated ones. Subsequently, no group has tested the assumption that hyophosphorylated pRb consists of multiply phosphorylated molecules by biochemically characterizing the extent or kinetics of pRb hypo-phosphorylation during G_1 phase. Given the importance of p16[INK4a] and cyclin D mutations in driving cancer progression [5, 15, 37], a precise understanding of how Cdk4/Cdk6 regulates pRb and thereby cell cycle progression would seem crucial to properly interpret the biological consequences of Cdk dysregulation and therapeutic inhibition.

A simple solution to directly test the nature of hypo-phosphorylated pRb is to quantitatively count phosphates on pRb during early G_1 phase when cyclin D-Cdk4/Cdk6 is active, an analysis that had never been performed on pRb. 1D SDS-PAGE and phospho-peptide mapping approaches cannot quantify phosphates on pRb; however, two-dimensional isoelectric focusing (2D IEF) can directly count the number of phosphates on full-length proteins. In 2D IEF, immunoprecipitated proteins are denatured in urea (no SDS), then resolved in the first dimension based on their isoelectric point (pI) by loading onto a fixed IEF pH strip from 3 to 10. When a protein hits its isoelectric point, it becomes neutral and precipitates into the IEF matrix at that location. In the second dimension, the proteins are resolved by molecular weight by soaking the IEF strip in SDS and loading it on top of a standard SDS-PAGE gel, followed by immunoblot to identify pRb (Fig. 6.3c) [31]. pRb has 14 Cdk consensus motifs (S/T-P-x-B) that are phosphorylated in vivo (two other Cdk sites, T5 and S567, are not phosphorylated) (Fig. 6.1). Phosphates are highly acidic modifications that decrease the overall pI of a protein with each additional phosphate added. Un-phosphorylated pRb has a pI ~8.2 and the addition of a single, mono-phosphate results in a pI ~7, whereas addition of 14 phosphates in hyper-phosphorylated pRb results in a pI <4.

Since pRb had not previously been subjected to IEF as far as we knew, we needed to calibrate the 2D IEF for detecting the number of phosphates on pRb. To do so, we generated pRb constructs where all Cdk sites were mutated to Alanine residues, termed ΔCdk pRb. We also generated mono-phosphorylation pRb constructs where we added back a single Cdk site, leaving the other 13 Cdk sites mutated as alanines. In addition, we added back three, six, nine, and all of the Cdk phosphorylation sites. 2D IEF of pRb from cycling cells expressing these phospho-constructs demonstrated for the first time the ability to quantitatively count the number of phosphates on pRb from 0 to 14 (Fig. 6.3d). Co-mixing of the ΔCdk pRb with endogenous pRb from quiescent T cells demonstrated these proteins to co-migrate consistent with un-phosphorylated pRb, confirming that pRb is only phosphorylated on Cdk sites in this system. Thus, this method   is ideal to accurately and reproducibly count the number of phosphates on pRb.

To determine the phosphorylation status of endogenous pRb during the cell cycle, we synchronized primary human fibroblasts by serum deprivation for 72 h, followed by serum stimulation, and took 12 time points over the next 16 h as cells passed from G_0 into early G_1 and late G_1 phases (Fig. 6.3e). Surprisingly, we found that pRb shifts from un-phosphorylated in G_0 to mono-phosphorylated species in early G_1 phase and remains exclusively mono-phosphorylated throughout the entire early G_1 phase. Importantly, we performed over 500 2D IEFs on 11 different cell types and always got the same results: pRb is exclusively mono-phosphorylated in early G_1 phase in all normal and p16[INK4a]-deficient tumorigenic cells. From the early to late G_1 transition, referred to as the restriction point by Pardee [32], mono-phosphorylated pRb undergoes a quantum conversion to hyper-phosphorylated pRb, displaying phosphorylation on all 14 Cdk sites (Fig. 6.3e). However, we found no biochemical evidence to support the notion of progressive, multiple hypo-phosphorylation of pRb during early G_1 phase. In other words, the notion of inter-mediate levels of phosphorylation of pRb vs. mono- and fully hyper-phosphorylated pRb, appears to be a theoretical notion only, leading to a rethinking of the function of cyclin D-Cdk4/Cdk6 in early G_1 phase.

Given the constitutive activity profile of cyclin D-Cdk4/Cdk6 throughout all of early G_1 phase (Fig. 6.3a), it arose as our top candidate for the pRb mono-phosphorylating kinase. To address this question, we combined synchronized cells and 2D IEF with three independent approaches to perturb Cdk4/Cdk6 kinase activity: (1) triple cyclin D knockout MEFs, (2) ectopic p16^{INK4a} expression, and (3) addition of Cdk4/Cdk6-specific inhibitors [31]. All three approaches gave the exact same answer, namely, that cyclin D-Cdk4/Cdk6 is the pRb mono-phosphorylating kinase. However, even more surprising was that cyclin D-Cdk4/Cdk6 generated 14 independent mono-phosphorylated pRb isoforms in early G_1 phase. We also deter-mined that the quantum inactivating hyper-phosphorylation of pRb at the late G_1 restriction point is performed by activation of cyclin E-Cdk2 complexes, not cyclin D-Cdk4/Cdk6. Together, these observations closely paralleled the physiologic activity profiles of Cdk4/Cdk6 with the E2F repression activity profile of pRb in early G_1 phase and derepression of E2F targets in late G_1 phase with Cdk2 activation and pRb hyper-phosphorylation. In contrast, the dogmatic view that cyclin D-Cdk4/Cdk6 complexes inactivate pRb through hyper-phosphorylation is difficult to recon-cile with the constitutive activity of Cdk4/Cdk6 throughout early G_1 phase.

6.2.3   Death of a Dogma Part III: Mono-phosphorylated pRb Is the Active Form of Rb in Early G1 Phase

The exclusive presence of 14 mono-phosphorylated pRb isoforms in early G_1 phase and the distinct absence of any higher order hypo-phosphorylated species raises the obvious question about the functional status of mono-phosphorylated pRb. Given that mono-phosphorylated pRb is the sole form of pRb present in early G_1 phase when E2F genes are repressed, by definition, some, most, or all of the 14 mono-phosphorylated pRb isoforms must be biologically active. pRb is a scaffold protein with no intrinsic enzymatic activity and has been shown to bind to over 100 pro-teins, including the prototypical binding to E2F1–4s, HDACs, and viral oncopro-teins E1a, TAg, and E7 [29]. We hypothesized that the generation of 14 mono-phosphorylated pRb isoforms may serve as a post-translational mechanism to diversify pRb from a single un-phosphorylated protein in G_0 to 14 independently functionalized mono-phosphorylated pRb isoforms that may each bind specific and likely overlapping cellular targets during early G_1 phase.

To test this hypothesis, we expressed un-phosphorylated ΔCdk pRb and each of the 14 mono-phosphorylated pRb constructs in cells co-expressing E1a or E2F-1, E2F-2, E2F-3, and E2F-4. Given its role in driving adenovirus infected quiescent G_0

cells (containing un-phosphorylated pRb) into early G_1 phase and then S phase, it was not too surprising that the E1a oncoprotein bound equally well to un-phosphorylated pRb and all 14 mono-phosphorylated pRb isoforms. This observation also showed that, by defnition, all single Cdk site mono-phosphorylated pRb proteins were correctly folded in vivo. Surprisingly, we found that all the mono-phosphorylated pRb isoforms avidly and preferentially bound to specific subsets of E2Fs. In other words, there was no mono-phosphorylated isoform of pRb that was unable to bind at least two species of E2F. While phosphorylation of T373 on pRb has been singled out as an inactivating phosphorylation site on a fragment of pRb [4], in our hands T373 mono-phosphorylated full-length pRb avidly bound to E1a, E2F1, E2F2, and E2F3. Together, these observations demonstrated that 14 independent mono-phosphorylated pRb isoforms that are present in early G_1 phase of 11 different normal and tumor cell types analyzed all differentially bound to E2F family members. We note that these observations parallel other signaling proteins where phosphorylation at specific sites enhances or depresses binding to cellular targets.

Although pRb regulates many processes in early G_1 phase [5], pRb's regulation of a DNA damage response-mediated cell cycle arrest is a critical function [2, 3, 14, 20]. Cells undergoing a DNA damage response activate cyclin D-Cdk4/Cdk6 complexes to generate active mono-phosphorylated pRb that regulates global transcription, whereas cells exiting the cell cycle to undergo differentiation use un-phosphorylated pRb [31]. Surprisingly, physiologic ectopic expression of mono-phosphorylated pRb in *RB1* deleted cells rescued the cell cycle arrest, regulated global transcription, and prevented the appearance of tetraploid cells during a DNA damage response, whereas un-phosphorylated ΔCdk pRb failed to rescue any of these phenotypes. However, in cells stimulated to exit the cell cycle to undergo differentiation, physiologic expression of un-phosphorylated ΔCdk pRb drove cells out of the cell cycle. Taken together, the new quantitative observations on pRb demonstrate that mono-phosphorylated pRb, generated by cyclin D-Cdk4/Cdk6 complexes, is the functionally active pRb isoform present in early G_1 phase, whereas un-phosphorylated pRb is the functionally active form in G_0 cells.

6.3   A New Working Model of G_1 Cell Cycle Progression and Lingering Questions

The new data that quantitatively counted the number of phosphates on pRb during early G_1 phase directly addressed several critical problems arising from numerous biochemical analyses of pRb phosphorylation going back more than 20 years. Most importantly, there is no biochemical evidence to support the concept that cyclin D-Cdk4/Cdk6 progressively hypo-phosphorylates pRb to produce a form of pRb bearing more than one phosphorylated Cdk recognition site. The exclusive presence of functionally active, mono-phosphorylated pRb generated by cyclin D-Cdk4/Cdk6 complexes in early G_1 phase significantly changes our understanding of G_1 cell cycle regulation. Incorporating the phosphorylation data into a new working model of G_1 cell cycle progression, we propose that G_0 cells contain functionally active, un-phosphorylated pRb that binds G_0 specific cellular targets and E2F1–4. Growth factor stimulation of cells entering early G_1 phase of the cell cycle induces D-type cyclins that in turn activate Cdk4/Cdk6 to convert un-phosphorylated pRb into 14 functionally active, mono-phosphorylated pRb isoforms that each are capable of binding to overlapping subsets of pRb target proteins, including binding to and repressing E2F target genes (Fig. 6.4). However, we view cyclin D-Cdk4/Cdk6 mono-phosphorylation of pRb as somewhat of a red herring for cell cycle progression in the sense that pRb mono-phosphorylation does not lead directly to activation of cyclin E-Cdk2 as the previous model predicted. Instead, we propose that cyclin D-Cdk4/Cdk6 complexes phosphorylate other critical (non-pRb family member) substrate(s) that drive a cellular metabolism pathway leading to activation of the Cdk-activating kinase (CAK) that activates cyclin E-Cdk2 complexes, resulting in pRb inactivation by hyper-phosphorylation, release of E2Fs, and transition across the restriction point into late G_1 and S phases (Fig. 6.4). Indeed, as described elsewhere in this volume, several candidate substrates of cyclin D-Cdk4/Cdk6 have been identified in recent years that could fulfill such a role.

Although the new working model incorporates modern quantitative biochemistry and cell biology, it leaves wide open several very significant questions. First and foremost, placing one and only one phosphorylation event on pRb at each of 14 sites by cyclin D-Cdk4/Cdk6 is very hard number to get to in biology. To our knowledge, there is no other example where the same phosphorylation sites are used singularly and multiply. Mechanistically speaking, how does cyclin D-Cdk4/Cdk6 transfer a single phosphate and then not transfer a second? We have been able to exclude phosphatases as the culprit [31]. We speculate that the N-terminal LxCxE domain on cyclin D binds the B box on pRb with a relatively low avidity [8] and that the presence of a single, highly negatively charged phosphate ionically repels cyclin D's LxCxE domain from rebinding, thereby preventing multi-phosphorylation. In contrast, cyclin E and A avidly bind pRb's four C-terminal R/KxL motifs outside of the pocket [1] (Fig. 6.1). Thus, cyclin E/A-Cdk2's strong binding to the C-terminal tail of pRb would allow access to all 14 Cdk sites on pRb even when transcription factors and chromatin-remodeling factors are bound to pRb's pocket and N-terminal binding sites. Because no multi-phosphorylated species of pRb were observed by 2D IEF, this would also explain the simultaneous switch-like inactivation of all 14 mono-phosphorylated pRb isoforms by one processive hyper-phosphorylation mechanism. This dramatic difference in avidity and binding location on pRb likely serves as the defining mechanism between pRb mono-phosphorylation and pRb hyper-phosphorylation.

There is no question that cyclin D-Cdk4/Cdk6 complexes are cancer drivers. So how is it possible that an oncogene is activating a tumor suppressor gene? Given that the majority of human tumors contain wild-type pRb, but select for deregulated cyclin D-Cdk4/Cdk6 activity [5, 7, 21, 36, 37], we hypothesize that the oncogenic activation of cyclin D-Cdk4/Cdk6 has two distinct consequences. First, pRb mono-phosphorylation serves to drive quiescent G_0 cells into an early G_1 phase phenotype (Fig. 6.4). By constitutively mono-phosphorylating pRb, the nascent

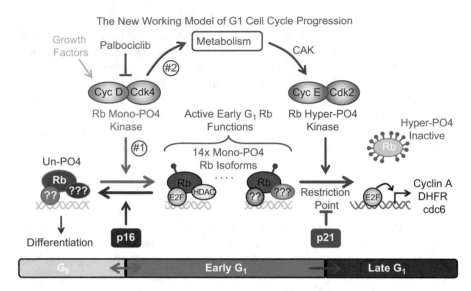

Fig. 6.4 The new working model of G_1 cell cycle progression. Un-phosphorylated pRb regulates G_0 cell cycle exit and differentiation. Growth factor signaling and DNA damage stimulate activation of cyclin D-Cdk4/Cdk6 complexes that diversify pRb into 14 mono-phosphorylated isoforms that independently bind specific cellular factors to regulate early G_1 phase functions and the DNA damage response. Cyclin D-Cdk4/Cdk6 mono-phosphorylation of pRb inactivates unphosphorylated pRb G_0 functions and thereby prevents cells from exiting the cell cycle. Activation of cyclin E-Cdk2 complexes at the late G_1 restriction point inactivates all 14 mono-phosphorylated pRb isoforms by a quantum hyper-phosphorylation. Cyclin A-Cdk2 and cyclin B-Cdk1 maintain pRb in an inactive hyper-phosphorylated state during S, G_2, and M phases. As cells complete cytokinesis, hyper-phosphorylated pRb is dephosphorylated by phosphatases and rapidly mono-phosphorylated by cyclin D-Cdk4/Cdk6 complexes. We speculate that an unknown metabolic sensor is upstream of cyclin E-Cdk2 activation. Deregulation of cyclin D-Cdk4/Cdk6 activity in cancer simultaneously inactivates un-phosphorylated pRb's G_0 functions and activates pRb's early G_1 phase functions by mono-phosphorylation, thereby driving cells from a low metabolism G_0 quiescence into a high metabolism early G_1 arrested state that prevents subsequent cell cycle exit or differentiation. We speculate that cyclin D-Cdk4/Cdk6 activity bifurcates and phosphorylates substrates in a metabolic pathway that ultimately converge on activation of the Cdk-activating kinase (CAK) leading to activation of cyclin E-Cdk2 and inactivation of pRb by hyper-phosphorylation

neoplastic cell avoids cell cycle exit and differentiation mediated by un-phosphorylated pRb. Second, what is the rate-limiting, switch-like mechanism that activates cyclin E-Cdk2, now seen as the first domino in pRb inactivation? We propose that constitutive cyclin D-Cdk4/Cdk6 phosphorylates key substrates in the metabolic pathway arm of the bifurcation that also likely requires additional oncogenic dysregulation of other signaling pathways but ultimately leads to premature activation of CAK and hence, activation of cyclin E-Cdk2 complexes and pRb hyper-phosphorylation [13, 31]. We are currently investigating the mechanics of this putative mechanism and the identity of the metabolic sensor.

6.4   Conclusion

In conclusion, the current view of the pRb pathway vis-a-vis regulation by cyclin D-Cdk4/Cdk6 was built without any quantitative biochemical data on the actual phosphorylation status of pRb during early G_1 phase, which allowed propagation of the incorrect notion of functional inactivation by cyclin D-Cdk4/Cdk6. Because p16^{INK4a} is a tumor suppressor regulating the oncogenic cyclin D-Cdk4/Cdk6 complexes that phosphorylate the pRb tumor suppressor, it was essentially guilt by association. However, the repression of E2F target genes during early G_1 phase never added up to fit the core tenet of the dogma. Current studies using quantitative biochemical analyses now make it clear that cyclin D-Cdk4/Cdk6 only mono-phosphorylates pRb to generate 14 independent active isoforms that bind to a wide variety of proteins, including E2Fs and viral oncoproteins. Functional inactivation of pRb by phosphorylation occurs at the hands of cyclin E-Cdk2, which hyper-phosphorylates pRb at the late G_1 checkpoint or restriction point. A significant question arising from this functional revision of the pRb pathway is how does cyclin E-Cdk2 become active? We predict that the answer will lead us to the missing pieces of G_1 cell cycle regulation and hopefully additional significant targets for cancer intervention.

References

1. Adams PD, Li X, Sellers WR, Baker KB, Leng X, Harper JW, Taya Y, Kaelin WG. Retinoblastoma protein contains a C-terminal motif that targets it for phosphorylation by cyclin-cdk complexes. Mol Cell Biol. 1999;19:1068–80.
2. Avni D, Yang H, Martelli F, Hofmann F, ElShamy WM, Ganesan S, Scully R, Livingston DM. Active localization of the retinoblastoma protein in chromatin and its response to S phase DNA damage. Mol Cell. 2003;12:735–46.
3. Brugarolas J, Moberg K, Boyd SD, Taya Y, Jacks T, Lees JA. Inhibition of cyclin-dependent kinase 2 by p21 is necessary for retinoblastoma protein-mediated G1 arrest after gamma-irradiation. Proc Natl Acad Sci U S A. 1999;96:1002–7.
4. Burke JR, Hura GL, Rubin SM. Structures of inactive retinoblastoma protein reveal multiple mechanisms for cell cycle control. Genes Dev. 2012;26:1156–66.
5. Burkhart DL, Sage J. Cellular mechanisms of tumour suppression by the retinoblastoma gene. Nat Rev Cancer. 2008;8:671–82.
6. Chen PL, Scully P, Shew JY, Wang JY, Lee WH. Phosphorylation of the retinoblastoma gene product is modulated during the cell cycle and cellular differentiation. Cell. 1989;58:1193–8.
7. Choi YJ, Anders L. Signaling through cyclin D-dependent kinases. Oncogene. 2014;33:1890–903.
8. Dowdy SF, Hinds PW, Louie K, Reed SI, Arnold A, Weinberg RA. Physical interaction of the retinoblastoma protein with human D cyclins. Cell. 1993;73:499–511.
9. Ewen ME, Sluss HK, Sherr CJ, Matsushime H, Kato J, Livingston DM. Functional interactions of the retinoblastoma protein with mammalian D-type cyclins. Cell. 1993;73:487–97.
10. Ezhevsky SA, Nagahara H, Vocero-Akbani AM, Gius DR, Wei MC, Dowdy SF. Hypophosphorylation of the retinoblastoma protein (pRb) by cyclin D:Cdk4/6 complexes results in active pRb. Proc Natl Acad Sci U S A. 1997;94:10699–704.

11. Ezhevsky SA, Ho A, Becker-Hapak M, Davis PK, Dowdy SF. Differential regulation of retino-blastoma tumor suppressor protein by G(1) cyclin-dependent kinase complexes in vivo. Mol Cell Biol. 2001;21:4773–84.

12. Gius D, Ezhevsky SA, Becker-Hapak M, Nagahara H, Wei MC, Dowdy SF. Transduced p16^{INK4a} Peptides Inhibit Hypo-Phosphorylation of the Retinoblastoma Protein and Cell Cycle Progression Prior to Activation of Cdk2 Complexes in Late G$_1$. Cancer Res. 1999;59:2577–80.

13. Haberichter T, Madge BR, Christopher RA, Yoshioka N, Dhiman A, Miller R, Gendelman R, Aksenov SV, Khalil IG, Dowdy SF. A systems biology dynamical model of mammalian G1 cell cycle progression. Mol Syst Biol. 2007;3:84–92.

14. Harrington EA, Bruce JL, Harlow E, Dyson N. pRB plays an essential role in cell cycle arrest induced by DNA damage. Proc Natl Acad Sci U S A. 1998;95:11945–50.

15. Henley SA, Dick FA. The retinoblastoma family of proteins and their regulatory functions in the mammalian cell division cycle. Cell Div. 2012;7:10.

16. Hinds PW, Mittnacht S, Dulic V, Arnold A, Reed SI, Weinberg RA. Regulation of retinoblas-toma protein functions by ectopic expression of human cyclins. Cell. 1992;70:993–1006.

17. Ho A, Dowdy SF. Regulation of G(1) cell-cycle progression by oncogenes and tumor suppres-sor genes. Curr Opin Genet Dev. 2002;12:47–52.

18. Johnson A, Skotheim JM. Start and the restriction point. Curr Opin Cell Biol. 2013;25:717–23.

19. Knudsen ES, Wang JY. Dual mechanisms for the inhibition of E2F binding to RB by cyclin-dependent kinase-mediated RB phosphorylation. Mol Cell Biol. 1997;17:5771–83.

20. Knudsen KE, Booth D, Naderi S, Sever-Chroneos Z, Fribourg AF, Hunton IC, Feramisco JR, Wang JY, Knudsen ES. RB-dependent S-phase response to DNA damage. Mol Cell Biol. 2000;20:7751–63.

21. Knudsen ES, Knudsen KE. Retinoblastoma tumor suppressor: where cancer meets the cell cycle. Exp Biol Med. 2006;231:1271–8.

22. Lamber EP, Beuron F, Morris EP, Svergun DI, Mittnacht S. Structural insights into the mecha-nism of phosphoregulation of the retinoblastoma protein. PLoS One. 2013;8:e58463.

23. Lee WH, Shew JY, Hong FD, Sery TW, Donoso LA, Young LJ, Bookstein R, Lee EY. The retinoblastoma susceptibility gene encodes a nuclear phosphoprotein associated with DNA binding activity. Nature. 1987;329:642–5.

24. Lee JO, Russo AA, Pavletich NP. Structure of the retinoblastoma tumour-suppressor pocket domain bound to a peptide from HPV E7. Nature. 1998;391:859–65.

25. Leng X, Connell-Crowley L, Goodrich D, Harper JW. S-Phase entry upon ectopic expres-sion of G1 cyclin-dependent kinases in the absence of retinoblastoma protein phosphorylation. Curr Biol. 1997;7:709–12.

26. Ludlow JW, DeCaprio JA, Huang CM, Lee WH, Paucha E, Livingston DM. SV40 large T antigen binds preferentially to an underphosphorylated member of the retinoblastoma suscep-tibility gene product family. Cell. 1989;56:57–65.

27. Lundberg AS, Weinberg RA. Functional inactivation of the retinoblastoma protein requires sequential modification by at least two distinct cyclin-cdk complexes. Mol Cell Biol. 1998;18:753–61.

28. Mittnacht S, Lees JA, Desai D, Harlow E, Morgan DO, Weinberg RA. Distinct sub-populations of the retinoblastoma protein show a distinct pattern of phosphorylation. EMBO J. 1994;13:118–27.

29. Morris EJ, Dyson NJ. Retinoblastoma protein partners. Adv Cancer Res. 2001;82:1–54.

30. Nagahara H, Ezhevsky SA, Vocero-Akbani A, Kaldis P, Solomon MJ, Dowdy SF. TGF-ß tar-geted inactivation of cyclin E:Cdk2 complexes by inhibition of Cdk2 activating kinase activity. Proc Natl Acad Sci U S A. 1999;96:14961–6.

31. Narasimha AM, Kaulich M, Shapiro GS, Choi YJ, Sicinski P, Dowdy SF. Cyclin D activates the Rb tumor suppressor by Mono-phosphorylation. elife. 2014. doi:10.7554/eLife.02872.

32. Pardee AB. A restriction point for control of normal animal cell proliferation. Proc Natl Acad Sci U S A. 1974;71:1286–90.

33. Paternot S, Bockstaele L, Bisteau X, Kooken H, Coulonval K, Roger PP. Rb inactivation in cell cycle and cancer: the puzzle of highly regulated activating phosphorylation of CDK4 versus constitutively active CDK-activating kinase. Cell Cycle. 2010;9:689–99.
34. Resnitzky D, Gossen M, Bujard H, Reed SI. Acceleration of the G1/S phase transition by expression of cyclins D1 and E with an inducible system. Mol Cell Biol. 1994;14:1669–79.
35. Sherr CJ. G1 phase progression: cycling on cue. Cell. 1994;79:551–5.
36. Sherr CJ, McCormick F. The RB and p53 pathways in cancer. Cancer Cell. 2002;2:103–12.
37. Sherr CJ, Beach D, Shapiro GI. Targeting CDK4 and CDK6: from discovery to therapy. Cancer Discov. 2016;6:353–67.
38. Yu B, Becker-Hapak M, Snyder EL, Vooijs M, Denicourt C, Dowdy SF. Distinct and non-overlapping roles for pRB and cyclin D:Cdk4/6 activity in Melanocyte survival. Proc Natl Acad Sci U S A. 2003;100:14881–6.

Index

© Springer International Publishing AG 2018   149
P.W. Hinds, N.E. Brown (eds.), *D-type Cyclins and Cancer*,
Current Cancer Research, DOI 10.1007/978-3-319-64451-6